HUMAN INTERFERON
Production and Clinical Use

ADVANCES IN EXPERIMENTAL MEDICINE AND BIOLOGY

Recent Volumes in this Series

Volume 101
ENZYMES OF LIPID METABOLISM
Edited by Shimon Gatt, Louis Freysz, and Paul Mandel

Volume 102
THROMBOSIS: Animal and Clinical Models
Edited by H. James Day, Basil A. Molony, Edward E. Nishizawa, and
Ronald H. Rynbrandt

Volume 103
HOMEOSTASIS OF PHOSPHATE AND OTHER MINERALS
Edited by Shaul G. Massry, Eberhard Ritz, and Aurelio Rapado

Volume 104
THE THROMBOTIC PROCESS IN ATHEROGENESIS
Edited by A. Bleakley Chandler, Karl Eurenius, Gardner C. McMillan,
Curtis B. Nelson, Colin J. Schwartz, and Stanford Wessler

Volume 105
NUTRITIONAL IMPROVEMENT OF FOOD AND FEED PROTEINS
Edited by Mendel Friedman

Volume 106
GASTROINTESTINAL HORMONES AND PATHOLOGY OF THE DIGESTIVE SYSTEM
Edited by Morton Grossman, V. Speranza, N. Basso, and E. Lezoche

Volume 107
SECRETORY IMMUNITY AND INFECTION
Edited by Jerry R. McGhee, Jiri Mestecky, and James L. Babb

Volume 108
AGING AND BIOLOGICAL RHYTHMS
Edited by Harvey V. Samis, Jr. and Salvatore Capobianco

Volume 109
DRUGS, LIPID METABOLISM, AND ATHEROSCLEROSIS
Edited by David Kritchevsky, Rodolfo Paoletti, and William L. Holmes

Volume 110
HUMAN INTERFERON: Production and Clinical Use
Edited by Warren R. Stinebring and Paul J. Chapple

HUMAN INTERFERON
Production and Clinical Use

Edited by

Warren R. Stinebring
The University of Vermont
Burlington, Vermont

and

Paul J. Chapple
W. Alton Jones Cell Science Center
Lake Placid, New York

PLENUM PRESS • NEW YORK AND LONDON

Library of Congress Cataloging in Publication Data

Interferon Workshop on Production of Human Interferon and Investigations of its Clinical
   Use, W. Alton Jones Cell Science Center, 1977.
   Human interferon: Production and clinical use.

   (Advances in experimental medicine and biology; v. 110)
   "Proceedings of the Interferon Workshop on Production of Human Interferon and In-
vestigations of its Clinical Use, held at the W. Alton Jones Cell Science Center, Lake Placid,
N Y., May 19–22, 1977."
   Includes index.
   1. Interferons—Congresses. 2. Interferons—Therapeutic use—Congresses. I. Stinebring,
Warren R. II. Chapple, Paul J. III. Title. IV. Series. [DNLM: 1. Interferon—Congresses. W1
AD559 v. 110/QW800 161p 1977]
   QR187.5.I572 1977   616.01'94   78-13393
   ISBN-13: 978-1-4615-9082-8   e-ISBN-13: 978-1-4615-9080-4
   DOI: 10.1007/978-1-4615-9080-4

Proceedings of the Interferon Workshop on Production of Human Interferon
and Investigations of its Clinical Use held at the W. Alton Jones Cell
Science Center, Lake Placid, New York, May 19–22, 1977

© 1978 Plenum Press, New York
Softcover reprint of the hardcover 1st edition 1978
A Division of Plenum Publishing Corporation
227 West 17th Street, New York, N.Y. 10011

PREFACE

PREFACE

This volume stems from a workshop held at the W. Alton Jones Cell
Science Center, Lake Placid, New York, May 19-20, 1977. The Center is an operational unit of the Tissue Culture Association and
offers, in collaboration with the Association's Education Committee, a wide range of activities. One of the earliest symposia
held at the Center was one concerned with Interferon, (Waymouth,
C. (Ed.) 1973. The Production and Use of Interferon for the Treatment and Prevention of Human Virus Infections. In Vitro, Monograph No. 3. Tissue Culture Association, Rockville, Maryland.) In
the intervening years since that symposium, many developments
have taken place. Much has been learned of the molecular biology
of interferon. However, the focus of this particular workshop,
as the title indicates, was the production of human interferon and
investigation of its clinical use.

Evidence has accumulated that interferon is an effective antiviral
and, possibly, an antitumor agent. The production of interferon
requires human cells and proper inducing agents. Present production facilities are limited. The quantities of interferon needed
for clinical trials and eventual routine demand the development of
new techniques in the growth and handling of cells and inducers.

Safety is an important consideration in the production of biological
products, and interferon is no exception. All components of the
system and the final product must satisfy stringent safety standards.

The participants of the workshop dealt with all of the above problems on a practical level. The next few years will see advances
in the technology of production of interferon. The future will
also see more widespread investigation of the antiviral and anti-
tumor roles of interferon. In addition, the lymphokine-like properties of interferon, the ability to affect cell growth and the
ability of interferon to affect lymphoid cells will all be investigated with the underlying purpose of utilizing interferon
in the treatment of human disease.

Thus, there will be an ever increasing need for larger and larger quantities of purified interferon. The editors and all those who took part in the workshop sincerely hope that this volume will act as a catalyst in bringing about the availability in sufficient quantity that it can be used for therapeutic purposes in a routine manner.

The editors wish to acknowledge the support and assistance of the Education Committee of the Tissue Culture Association and the financial support of the Bureau of Biologics of the Food and Drug Administration, National Institute of Allergy and Infectious Diseases, National Cancer Institute, and the Fogarty International Center.

This volume would not have been possible without the hard work and perseverance of Mrs. Ellen Anastos-Dorchak, who kept the editors on track and who supervised the typing of all the manuscripts in a form ready for publication.

CONTENTS

Large Scale Production and Properties of Human Leukocyte
        Interferon Used in Clinical Trials   1
    H.-L. Kauppinen, G. Myllylä, and K. Cantell

New Microcarriers for the Large Scale Production of
        Anchorage-Dependent Mammalian Cells   15
    D.W. Levine, W.G. Thilly, and D.I.C. Wang

Effect of Low Levels of Cyclic Ribonucleotides on Mitogen
        and Virus Induced Production of Interferon   25
    H.M. Johnson and S. Baron

Production of Interferon in Human Cell Cultures by a New,
        Potent Viral Inducer   37
    P. Jameson and S.E. Grossberg

Human Interferon: Large Scale Production in Embryo
        Fibroblast Cultures   55
    V.G. Edy, J. VanDamme, A. Billiau, and P. DeSomer

Factors Influencing Production of Interferon by Human
        Lymphoblastoid Cells   61
    M.D. Johnston, K.H. Fantes, N.B. Finter, and B. Chir

Antigenic Properties and Heterospecific Antiviral
        Activities of Human Leukocyte Interferon Species. .   75
    K. Paucker, B.J. Dalton, and E.T. Torma

Membrane Alterations Following Interferon Treatment   85
    E.H. Chang, E.F. Grollman, F.T. Jay, G. Lee, L.D. Kohn,
    and R.M. Friedman

Selection of New Human Foreskin Fibroblast Cell Strains
        for Interferon Production   101
    J. Vilcek, E.A. Havell, M.L. Gradoville, M. Mika-Johnson,
    and W.H.J. Douglas

Tissue Culture Models of In Vivo Interferon Production
and Action 119
F. Dianzani, I. Viano, M. Santiano, M. Zucca,
P. Gullino, and S. Baron

Thermal and Vortical Stability of Purified Human Fibroblast
Interferon 133
J.J. Sedmak, P. Jameson, and S.E. Grossberg

Interferon Assay Anomaly Variation of Interferon Response
with Cell Type and Sialic Acid Content 153
A.A. Schwartz and D. Villani-Price

Interferon Therapy for Neoplastic Diseases in Man In Vitro
and In Vivo Studies 159
S. Einhorn and H. Strander

Results of a Five-Year Study of the Curative Effect of Double
Stranded Ribonucleic Acid in Viral Dermatoses and
Eye Diseases 175
L. Borecky, J. Buchvald, E. Adlerova, I. Stodola,
E. Obrucnikova, Z. Gruntova, V. Lackovic, and
J. Doskocil

The Possibility of Interferon Production in Tumors 193
W.R. Stinebring and S. Jenkins

The Future of Interferon as an Antiviral Drug 201
J.K. Dunnick and G.J. Galasso

Considerations in Our Search for Interferons for Clinical
Use 213
R.Z. Lockart, Jr. and E. Knight

Index 217

LARGE SCALE PRODUCTION AND PROPERTIES OF HUMAN LEUKOCYTE

INTERFERON USED IN CLINICAL TRIALS

H.-L. Kauppinen, M.Sc., G. Myllylä, M.D.,
and K. Cantell, M.D.

Finnish Red Cross Blood Transfusion Centre and
Central Public Health Laboratory, Helsinki, Finland

INTRODUCTION

Human leukocyte interferon has been produced for clinical use
in our laboratories for several years (1). A routine procedure
for the production and purification has been established (2,3),
but studies to improve the recovery and quality of the clinical
interferon continue. Since the availability of fresh leukocytes
is the factor limiting production, maximization of the interferon
yield per blood unit is of prime importance.

In clinical trials with interferons some side-effects including
pyrogenicity have appeared. These have prompted us to analyse the
impurities of the interferon preparations to see whether the side-
effects could be eliminated by further purification. This paper
describes our recent studies on the production of the human
leukocyte interferon for clinical use and on the properties of the
interferon preparations employed in the current clinical trials.

PRODUCTION

Storage of Blood

Routinely, only buffy coats from fresh blood are collected.
In practice the collection takes place 2-5 hours after donation.
The buffy coats are stored overnight at 4°C and used the next day
for interferon production. If production is started immediately
after harvesting the cells, the interferon titers are only slightly
higher. If the buffy coats are harvested from one-day-old blood and
in addition stored overnight so that the total storage time is two

1

days, the interferon titers are clearly reduced.   With additional
storage the titers are reduced further (Fig. 1).

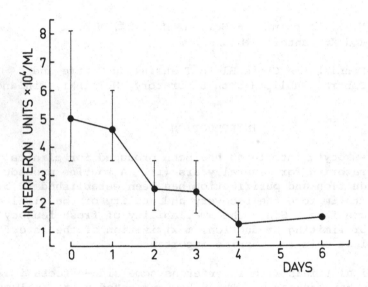

Figure 1

Production of interferon by leukocytes stored for different periods
of time at 4°C.

When the blood is stored before harvesting the buffy coat, the
leukocytes separate in centrifugation as a more distinct layer
and more leukocytes can be recovered from the same volume.   The
maximum yield of leukocytes counted after purification of the cells
is gained when the blood is stored for one day before harvesting

the buffy coat (Fig. 2). As in our routine procedure the cells are
stored overnight, so that the total storage time is two days. With
longer storage the aggregation of the cells reduces the final yield
of viable cells. Altogether the total recovery of interferon units
per buffy coat is not essentially affected by the storage of the
blood or the leukocytes for 1-2 days at 4°C.

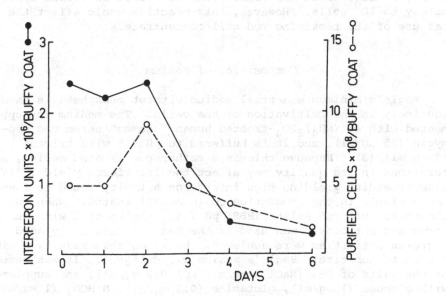

Figure 2

Recovery of leukocytes (cells/buffy coat 0———0) and interferon
(units/buffy coat ●———●) in relation to the storage of the blood.

Recovery of Leukocytes

Our routine way of collecting leukocytes is to press about
13 ml buffy coat from fresh blood after centrifugation of the bag.
Such buffy coat contains on average 0.95 x 10^9 leukocytes. One
blood unit of 500 ml CPD-blood contains 3.0 x 10^9 leukocytes (4).
Thus only about one third of the leukocytes are recovered. During
the two purification steps of the cells (1,3) about a half of them
are lost. The final leukocyte yield is 0.47 x 10^9 cells per buffy
coat of 12.3 ml (mean of 8815 buffy coats). It seems that the
lost cells are disintegrated during the process, since it has not
been possible to recover the leukocytes in any of the waste solutions.

There are ways of increasing the recovery of leukocytes in
collection from fresh blood. If the sharp-cornered bag used
routinely is replaced by a cone-shaped plastic blood bag the
leukocyte yield should increase considerably, because the lodge-
ment of cells in the corners of the bag during the pressing process
is precluded. On the other hand by harvesting a much larger volume,
40 ml, of buffy coat the final leukocyte yield increases approximately to 10^9 cells. However, this practice would affect the clinical use of the remaining red cell concentrate.

Composition of Medium

Eagle's minimum essential medium without phosphate is used
routinely in the cultivation of the cells. The medium is supplemented with 4% $(NH_4)_2SO_4$-treated human "agamma" serum and neomycin (25 µg/ml), and it is buffered to pH 7.5 with tricine
(3 mg/ml) (3). Because this is a rather complicated medium, alterations in its quality may affect the interferon yields. If a
simpler medium yielding high interferon activity could be found,
one variable in the production could be eliminated. When the
phosphate buffered saline (PBS) pH 7.4, supplemented with human
serum and antibiotic was used as the medium, relatively good interferon activities were achieved. Based on this result, a simpler
medium to substitute Eagle's medium was designed. It was based
on the salts of PBS (NaCl 8 mg/ml; KCl 0.2 mg/ml), and supplemented
with glucose (1 mg/ml), glutamine (0.3 mg/ml), $NaHCO_3$ (1 mg/ml),
4% human "agamma" serum and neomycin (25 µg/ml), buffered with
tricine (3 mg/ml) to pH 7.5. This medium proved to have the same
slightly hypotonic osmolality of 268 mOsm/kg as Eagle's medium.

Interferon was produced in 24 batches, in total 65 liters in
this enriched saline-medium, parallel with 42 batches, total 108
liters, in Eagle's medium. The mean interferon titers in both
media were of the same order of magnitude, namely 34400 units/ml
in the saline-medium and 45000 units/ml in Eagle's medium. Because

of the small difference we still routinely use Eagle's medium.

Storage of Sendai-virus

The way of storing the Sendai-virus used for interferon induction has proved to be very critical. The fresh virus can be
stored at +5°C for weeks without supplements. If longer storage
is needed, the virus can be stored frozen at -70°C, but then an
addition of 4% "agamma" serum is necessary to maintain the interferon inducing capacity. After thawing, the virus must be used
within one week.

PROPERTIES

Acid Treatment of Crude Interferon

Purification of the crude leukocyte interferon is routinely
done by the ethanol fractionation method (1-3). After adopting
this method, inactivation of the inducing Sendai-virus by acid
treatment was abandoned, because the inactivation is achieved by
the ethanol.   Only those batches of crude interferon which have
been contaminated by bacteria have been stored at pH 2 for five
days to stop the bacterial growth.

Twenty purification batches of 21.5 liters each in which crude
interferon was acid treated for five days were purified in parallel
with 29 ordinary batches (Table 1). Regardless of the pretreatment,
the mean recovery was approximately 40%. On the other hand, the
purity of the end product was clearly affected by acid treatment.
The mean specific activity of the partially purified interferon
was approximately 2.4 times higher in the non-acid treated material
compared with that of the acid treated interferon. The cellulose
acetate electrophoresis pattern (Fig. 3) shows more heterogeneity of
impurities when the starting material for P-IF has been pH 2 treated.

Table 1

The effect of acid treatment of crude interferon on the purity of
the P-IF preparations obtained by the ethanol fractionation method
(1-3).

Treatment	Number of experiments	Mean recovery%	Mean specific activity units/mg
None	29	39.5	7.94×10^5
pH 2,5 days	20	42.8	3.29×10^5

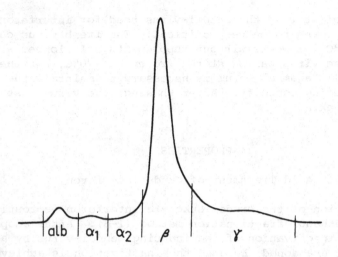

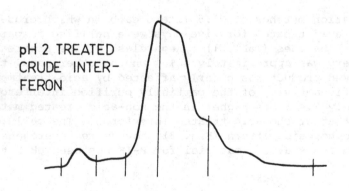

pH 2 TREATED
CRUDE INTER-
FERON

Figure 3

Electrophoresis of PI-F derived from non-treated and pH 2-treated
crude interferon on cellulose acetate at pH 8.6.

Pyrogenicity

The interferon concentrates and some P-IF preparations of lower purification grade have been found to cause pyrogenic reactions in patients (6). A study was made to see whether the pyrogenicity could be eliminated or diminished by further purification. The fever reactions in rabbits were compared with a limulus endotoxin pyrogen assay (Pyrogent^R, Mallincrodt, Inc., St. Louis).

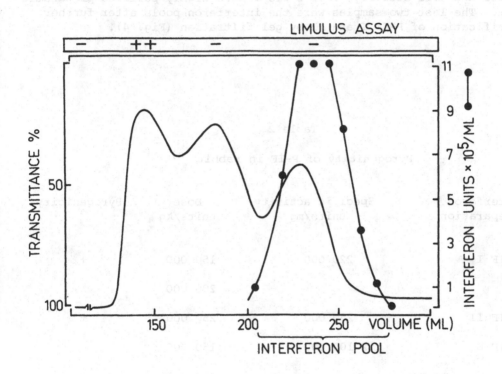

Figure 4

Purification of P-IF by gel filtration on Sephadex G-100. A sample of 12 ml P-IF, 6 million units/ml, 220 000 units/mg protein, was run on a 4.5 x 88 column in phosphate buffered saline, pH 7.4. The sample was positive with the limulus assay.

A P-IF preparation with a fairly low specific activity of 220000 units/mg was fractionated by gel filtration on a Sephadex G-100 column (Fig. 4). The interferon activity was detected at the last recorded peak. This preparative scale fractionation gives about 5-10-fold additional purification of the interferon. The column set and the buffers were all tested to be pyrogen free by the

limulus assay. The highly concentrated sample of P-IF (protein
content 27 mg/ml) gave a positive limulus reaction. After fractionation by Sephadex G-100 only the pool around the first detected peak
was posivite. The interferon pool was always negative even after 10-
fold concentration by lyophilization.

The pyrogenicity of five different P-IF samples of three types wa
tested in rabbits (Table 2). P-IF I and II represent preparations of
lower specific activity, P-IF B was a more highly purified preparatic
(2). The last two samples were the interferon pools after further
purification of P-IF I and II by gel filtration (Fig. 4).

Table 2

Pyrogenicity of P-IF in rabbit [a]

Interferon preparation	Specific activity units/mg	Dose units/kg	Pyrogenicity
P-IF I	220 000	150 000	+
		200 000	±
P-IF II	250 000	250 000	+
P-IF B	2 100 000	150 000	−
		250 000	±
		350 000	+
Sx IF-pool (I)	1 000 000	150 000	−
Sx IF-pool (II)	2 200 000	250 000	+

[a] The test was performed in parallel in three rabbits. The dose
(units/kg) is equivalent to the sample dilution (units/ml). One
milliliter was injected intravenously per kg of rabbit weight. The
summed rise of temperature in rabbits followed during 5 hours was
measured. 1.1°C was the limit of pyrogenicity. All samples were
negative with the limulus test.

The dose of interferon in the rabbits was quite high, corresponding to about 10-20 million units in an adult human, but with
smaller doses the test was always negative. All samples started
to show pyrogenicity at a dose of around 200 000 units per kilogram
of rabbit weight. The pyrogenicity in the rabbits did not correlate
with the purity of the sample. This finding does not exclude the
possibility that interferon itself could be pyrogenic.

Protein Composition

A sample of P-IF (220 000 units/mg) and the corresponding
Sephadex G-100 purified interferon pool (Sx IF-pool) (Fig. 4) were
analysed by immunological precipitation with rabbit anti normal
human serum (NHS) and rabbit anti Sx IF-pool serum and with the
same sera after selective adsorption by human serum proteins. The
analysis was performed by crossed immunoelectrophoresis in agarose
(7). The results showed (Fig. 5) that P-IF contained monomeric
and polymeric albumin, two unidentified serum proteins and two
proteins antigenically not identical with serum proteins (precipi-
tating with NHS-adsorbed anti Sx IF-pool serum). Further, one more
protein was detected which did not move in the electrical field
but precipitated with anti NHS. It probably represented some serum
proteins denatured due to the ethanol treatment (1-3). The more
purified preparation, Sx IF-pool, still contained the same proteins
except for the polymeric albumin and the denatured proteins. The
proteins of the Sendai virus could not be detected with the anti
Sx IF-pool serum.

Sterilization of interferon by membrane filtration

Many interferons are known to lose activity during sterilization
by membrane filtration (5). We have used a standard cellulose ester
bacterial filter (MF-Millipore, GS, diameter 47 mm, Millipore Corp.,
Bedford, Mass.) for filtration experiments with P-IF. The results
depend on the protein content of the interferon preparation. If
the protein content is in the order of some micrograms per ml a
marked adsorption of interferon of the membrane is found at the
beginning of the filtration, but full recovery is obtained upon
further filtration. The initial loss can be avoided by saturating
the membrane with albumin (10 ml, 5% human serum albumin) or by
adding albumin (0.5%) to the interferon samples.

The interferon activity and the protein content of the P-IF
preparations used in the clinical trials are so high (ca. 5×10^6
units/ml, 5-10 mg protein/ml) that such preparations can be filtrated
through MF-Millipore membrane without any significant loss of
activity.

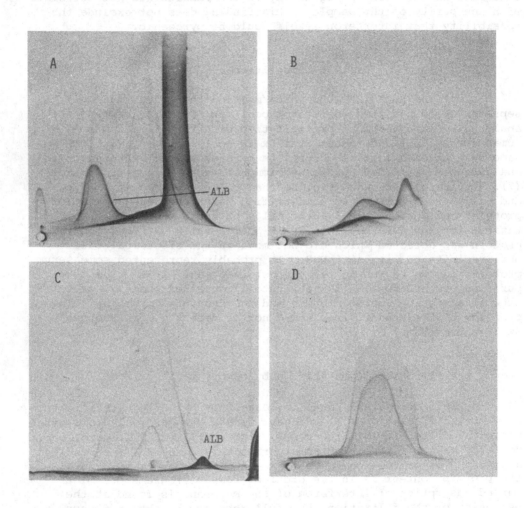

Figure 5

Crossed immunoelectrophoresis in agarose (7) of PI-F preparations.
P-IF against anti NHS serum (A) and NHS-adsorbed anti Sx IF-pool
serum (B), Sx IF-pool against anti NHS serum (C) and NHS-adsorbed
anti Sx IF-pool serum (D).

Stability

The stability of crude interferon (Fig. 6) and P-IF (Fig. 7) was followed during long storage periods at different temperatures. Both preparations had a pH of 7.4 and did not contain any additives. The results show that the crude interferon could be stored for about one year at 5°C without marked loss of activity. The P-IF maintained full activity for at least one year when stored at 5°C. Likewise the activity was maintained when the material was stored frozen at -70°C. But a room temperature the activity was markedly decreased after one year. Some precipitate was observed at the bottom of the test tubes. At least a part of missing activity could be recovered from this precipitate.

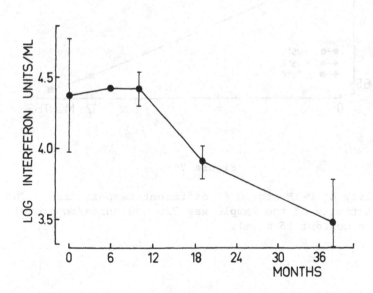

Figure 6

The stability of crude interferon at 4°C. The specific activity of the sample was 10 000 units/mg protein and the protein content 2.4 mg/ml. Each point is the mean of at least 4 assays (8).

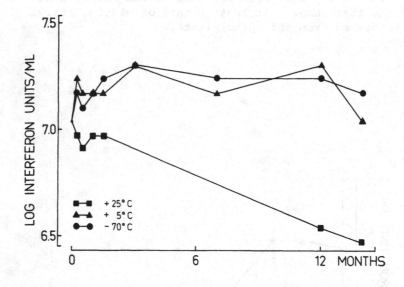

Figure 7

The stability of P-IF stored at different temperatures. The specific activity of the sample was 720 000 units/mg protein and the protein content 15 mg/ml.

References

1. Cantell, K., Hirvon, S., Mogensen, K.E., and Pyhälä, L. In: In Vitro, Ed. C. Waymouth. The Tissue Culture Association, Rockville, 1974, pp. 35-38.
2. Cantell, K., and Hirvonen, S. Texas Rep. Biol. Med., In press.
3. Mogensen, K.E., and Cantell, K. Pharmacology and Therapeutis, In press.
4. Tenczar, F.J. Transfusion 13: 183, 1973.
5. Shinkai, K., and Nishimura, T. Japan. J. Microbiol. 20: 251, 1976.
6. Strander, H., Cantell, K., Carlström, G., and Jacobson, P. J. Nat. Cancer Inst. 51: 733, 1973.
7. Weeke, B. Scand. J. Immunol., Vol. 2, Suppl. 1, 1973, pp. 47.
8. Strander, H., and Cantell, K. Ann. Med. Exp. Fenn. 44: 265, 1966.

NEW MICROCARRIERS FOR THE LARGE SCALE PRODUCTION OF ANCHORAGE-DEPENDENT MAMMALIAN CELLS

D.W. Levine, W.G. Thilly, and D.I.C. Wang

Department of Nutrition and Food Science, Massachusetts

Institute of Technology, Cambridge, Massachusetts

INTRODUCTION

One major limitation in the use of mammalian cells for the production of a product such as interferon is the technology for obtaining cells in sufficient quantity to meet the demands of the product's usage. For the growth of surface-dependent mammalian cells, this problem translates into one of providing sufficient total surface area, and providing that surface in a configuration convenient for large scale process. The standard growth procedures using dishes and roller bottles are sufficient to meet the small scale demands of laboratory work, however, the scale up process for these systems presents many difficulties. Since each roller bottle has limited production capability, scale up is achieved by increasing the number of bottles used. This multiplicity of low productivity reactors is physically difficult to manipulate (e.g. for medium changes, or cell harvest) and virtually impossible to maintain in a homogeneous state.

A number of alternative methods of cell culture have been proposed. These techniques include: plastic bags (Munder et al., 1971); stacked plates (Weiss and Schleicher, 1968); roller bottles having interior surfaces of spiralled plastic film (House et al., 1972); a packed bed of glass beads (Wohler et al., 1972); artificial capillaries (Knazek et al., 1972) and the microcarrier technique (van Wezel, 1967, 1972, 1973). These methods have been reviewed by Litwin (1971) and Maroudas (1973). Since the ultimate limitation of cell growth is the surface area available, the ability of a vessel to produce cells, that is, its productivity (cells/liter-hour) will depend on the total Surface per Volume, or S/V. Since we are attempting to maximize productivity, we can compare these techniques

for current S/V and for potential increases in S/V. Also since
our eventual goal is to apply these technologies to large scale
procedures, we may consider the scale up potential of these
various techniques. Estimates for these data are presented in
Table I. The microcarrier technique performs well on the basis
of each of these criteria.

Table I

A Comparison of S/V Value Estimates for Cell Propagators

System	S/V (cm^2/cm^3)	Reference
Roller Bottles	.2-.7	
Plastic Bags	5.0	Munder et al., 1971
Multiplate Propagator	1.7	Weiss and Schleicher, 1968
Spiral Film	4.0	House et al., 1972
Glass Bead Propagator	10.0	Wohler et al., 1972
Artificial Capillary	31	Knazek et al., 1972
Microcarriers	36	van Wezel, 1973

	Potential S/V Increase	Potential Physical Scale-up
Roller Bottles	-	-
Plastic Bags	$\pm(\leq 2$ fold)	-
Multiplate Propagator	-	+
Spiral Film	$\pm(\leq 2$ fold)	+
Glass Bead Propagator	-	-
Artificial Capillary	-	$\pm$
Microcarriers	$+(\leq 10$ fold)	+

BACKGROUND

The microcarrier system was first described by van Wezel and his coworkers (1967, 1972, 1973, and van Hemert et al., 1969). In his work, he utilized a commercially available anion exchange resin (DEAE-Sephadex A50) as a support material. With this material, they demonstrated the growth of numerous surface-dependent cell types (primary, normal, and continuous lines), and they documented these cell types' abilities to produce viruses. However, there are certain drawbacks associated with the DEAE-Sephadex A50 system which has tended to limit the acceptability of the microcarrier technique. Specifically, for the growth of primary and normal cell strains, these microcarriers tend to destroy a significant portion of the inoculum, induce long lag times, and inhibit culture growth (van Wezel, 1967, 1972, 1973; Horng and McLimans, 1975; Levine et al., 1975). Furthermore, the severity of these effects is increased with increasing microcarrier concentration. Since the total surface area per volume of the culture will be proportional to the micro-carrier concentration, these toxic effects have the overall result of limiting the usable surface area per volume and hence limiting cell productivity. Various attempts have been made by workers in the field to circumvent these toxicities (van Wezel, 1973; van Hemert, 1969; Horng and McLimans, 1975; Levine et al., 1975), but these attempts have failed to produce a microcarrier system totally free of the limitations noted above.

EXPERIMENTAL RESULTS

Our initial efforts (Levine et al., 1975) to reduce the apparent toxicities of the commercially available material were based on the hypothesis that the observed toxicities could be the result of the binding of essential nutrients to the anion exchange sites on the microcarriers, thus making them unavailable to the cells. To counteract this adsorption, we added a non-toxic polyanion (car-boxymethylcellulose, CMC) to the growth medium. We found that al-though this treatment tended to protect the inoculum and reduce the initial destruction of cells, the microcarrier system stilled showed growth inhibition at higher microcarrier concentrations, and still showed the same limited saturation cell concentrations as in the use of untreated DEAE-Sephadex A50. That is, while the CMC addition successfully protected cell integrity, allowing increases in microcarrier concentration, the saturation cell concentration was still severely limited, even in the presence of excess surface area, and excess nutrients.

An alternate approach to optimizing microcarrier culture growth was to synthesize our own microcarrier, carefully controlling the degree of substitution with anion exchange sites (Methods and Pro-cedures are documented in Levine et al., 1977). For these studies,

we used a microcarrier material similar to the commercially available DEAE-Sephadex A50. In this case, however, we bound varying
amounts of diethylaminoethyl-(DEAE) to a Sephadex G-50 matrix
(uncharged crosslinked dextran). By screening a series of microcarriers at varying degrees of substitution (milliequivalents of
Cl^- per gram of dry, unreacted G-50) we established that cell growth
is optimal in the range of 1.5-2.5 meq/gram. This capacity should
be compared to that of the commercially available DEAE-Sephadex A50,
at >6.0 meq/gram unreacted dextran. Our data suggest that the adverse
effects of the DEAE-Sephadex A50 are not related to nutrient uptake,
but rather to the high degree of substitution with the DEAE-moiety.

Microcarriers of optimal DEAE substitution gave excellent
growth results when used in culturing normal diploid human fibroblasts, HEL 299 (ATCC #CCL 137) (in DMEM plus 10% fetal calf serum),
and culturing secondary chicken embryo fibroblasts (in DMEM plus
1% calf serum, 1% chicken serum, and 2% tryptose phosphate broth).
Figure 1 compares the growth behavior of HEL 299 on the commercially
available carrier (DEAE-Sephadex A50) with the behavior of the reduced charge capacity carriers synthesized in our laboratory, at
carrier concentraitons of 2 gram unreacted dextran/liter. As
shown the new carrier material shows no tendency to destroy the
inoculum, does not cause significant lag times, and allows for
culture growth up to a high cell concentration of 1.4×10^6 cells/ml
More importantly, the carrier concentration may be increased to
5 gram unreacted dextran/liter with a concommitant increase in
saturation cell concentration, as shown in Figure 2, for the growth
of normal diploid human fibroblasts, and secondary chicken fibroblasts. Even at high microcarrier concentrations there is again
no evidence of those toxicities or limitations encountered with
the commercially available materials. Culture saturation cell
concentrations of $3.0-4.0 \times 10^6$ cells have been reproducibly obtained using this technique. At this point we feel that factors
other than the microcarrier surface and limiting cell concentration.

To put these numbers into perspective, we may compare cell
growth in microcarrier culture to cell growth in roller bottles
(Corning, 490 cm^2). In our hands, HEL 299 normal diploid human
fibroblasts grew typically to 5×10^7 cells 490 cm^2 roller bottle.
Using this figure and a microcarrier saturation cell concentration
of 3.0×10^6 cells/ml, we find that a one liter microcarrier culture
is equivalent to 60 roller bottles. This comparison becomes quite
striking when one considers that a moderately sized growth vessel
of 100 liters is the growth equivalent to 6000 roller bottles. A
similar comparison utilizing growth data for the interferon producing FS-4 strain, yield comparative numbers of 30-50 roller
bottles (490 cm^2) per one liter microcarrier culture.

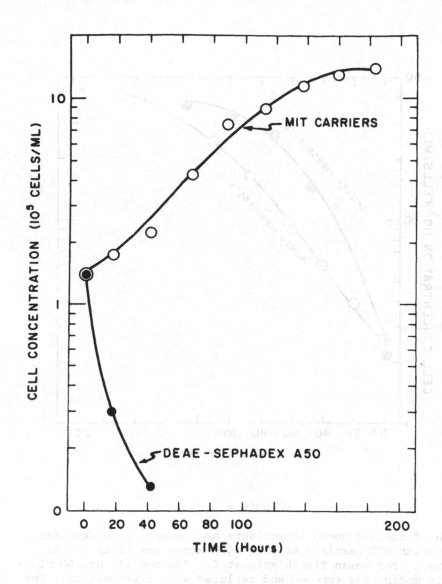

Figure 1

Growth of diploid human fibroblasts on MIT-carriers and DEAE
Sephadex A-50 at carrier concentrations of 2 mg of dextran per ml
($\sim$13 cm^2/ml).

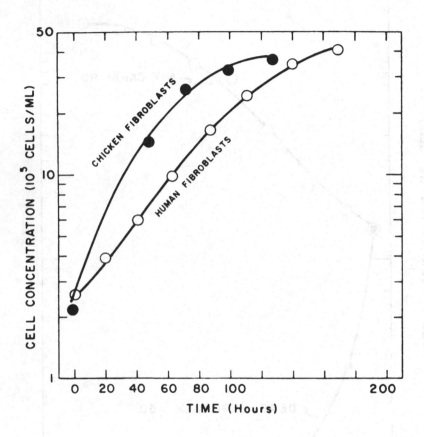

Figure 2

Growth of diploid human fibroblasts and secondary chicken fibroblasts with MIT-carriers at 5 mg of dextran per ml in 100 ml
cultures. For human fibroblasts at 65, 100 and 120 hr, 50 ml of
culture medium was removed and replaced with fresh medium. For
secondary chicken fibroblasts at 50, 75 and 100 hr, 50 ml of
culture medium was removed and replaced with fresh medium. No
cells were removed in this process.

The development of this new microcarrier eliminates many of the previously existing objections to the application of microcarrier culture to large scale growth of surface dependent cells. These developments have eliminated the apparent toxicities observed with other microcarrier systems, and allow for culture growth to high cell concentrations.

We are currently extending our work to the growth of other cell strains, lines and types. We are examining the growth of viruses in microcarrier grown cells. And we continue to examine the nature of the microcarrier material to optimize the chemistry of the microcarrier configuration.

ACKNOWLEDGEMENTS

I would like to acknowledge Jason S. Wong for his contribution to the development of the new microcarrier.

This research was supported in part by the National Science Foundation Grant # BMS 74 5676 A01. David W. Levine was supported in part by a Health Education and Welfare NIH Training Grant # 5T01ES 00063, and by the MIT Health Science Fellowship Program.

REFERENCES

1. Horng, C. and W. McLimans. 1975. Primary suspension culture of calf anterior pituitary cells on a microcarrier surface. Biotechnol. Bioeng. 17:713.

2. House, W., M. Shearer and N.G. Maroudas. 1972. Method for bulk culture of animal cells on plastic films. Exp. Cell Res. 71:293.

3. Knazek, R.A., P.M. Gullino, P.O. Kohler and R.L. Dedrick. 1972. Cell culture on artificial capillaries: An approach to tissue growth in vitro. Science 178:65.

4. Levine, D.W., D.I.C. Wang and W.C. Thilly. 1975. Optimizing parameters for growth of anchorage-dependent mammalian cells in microcarrier culture. In R.T. Acton and J.D. Lynn, eds. Presented at Cell Culture Congress, Birmingham, Alabama. Reprinted in Cell Culture and Its Application. Academic Press, New York, 1977.

5. Levine, D.W., J.S. Wong, D.I.C. Wang and W.G. Thilly. 1977. Microcarrier cell culture: New methods for research-scale application. Somatic Cell Genetics 3:149.

6. Litwin, J. 1971. Mass cultivation of mammalian cells. Process Biochemistry 6:15.

7. Maroudas, N.G. 1973. New methods for large-scale culture of anchorage-dependent cells. In R.H. Pain and B.J. Smith, eds. New Techniques in Biophysics and Cell Biology. John Wiley and Sons, New York.

8. Munder, P.G., M. Modolell and D.F.H. Wallach. 1971. Cell propagation on films of polymeric fluorocarbon as a means to regulate pericellular pH and pCO_2 in cultured monolayers. FEBS Letters 15:191.

9. van Hemert, D., D.G. Kilburn and A.L. van Wezel. 1969. Homogeneous cultivation of animal cells for the production of virus and virus products. Biotechnol. Bioeng. 11:875.

10. van Wezel, A.L. 1967. Growth of cell-strains and primary cells on microcarriers in homogeneous culture. Nature 216:64.

11. van Wezel, A.L. 1972. New trends in the preparation of cell substrates for the production of virus vaccines. Prog. Immunobiol. Standard. 5:187.

12.   van Wezel, A.L.   1973.   Microcarrier cultures of animal cells. Pg. 372 in P.F. Kruse and M.K. Patterson, eds.   Tissue Culture, Methods and Applications.   Academic Press, New York.

13.   Wohler, W., H.W. Rudiger and E. Passarge.   1972.   Large scale culturing of normal diploid cells on glass beads using a novel type of culture vessel.   Exp. Cell Res. 74:571.

14.   Weiss, R.E. and J.B. Schleicher.   1968.   A multisurface tissue propagator for the mass-scale growth of cell monolayers.   Biotechnol. and Bioeng. 10:601.

EFFECT OF LOW LEVELS OF CYCLIC RIBONUCLEOTIDES ON MITOGEN AND VIRUS

INDUCED PRODUCTION OF INTERFERON

Howard M. Johnson and Samuel Baron

Department of Microbiology
University of Texas Medical Branch
Galveston, Texas   77550

In the murine system there are at least two distinct types
of interferons. One type (virus-type interferon) is induced by
viruses and polyribonucleotides (Isaacs, 1963; Field et al., 1967),
while the other type (immune interferon) is produced, along with
other mediators by antigen stimulation of sensitized lymphocytes
(Green et al., 1969; Salvin et al., 1973) and by T lymphocyte
mitogens (Wheelock, 1965; Stobo et al., 1974; Johnson et al., 1977).
The interferons are antigenically distinct (Youngner et al., 1973;
Johnson et al., 1976). A similar classification of human interferons exists (Havell et al., 1975).

The data presented here will concentrate on immune interferon
induction, in terms of the mitogens most suitable for optimal
induction, and the regulatory role of adenosine 3', 5'-cyclic monophosphate (cAMP) in this induction. Data on immune interferon
induction by T cell mitogens and the relationship of this to cAMP
effects on suppressor and helper cell activities in the in vitro
antibody response will also be presented. These studies provide
useful models for studies on immune interferon induction by
peripheral human lymphocytes.

The relative abilities of various concentrations of three
T cell mitogens, concanavalin A (Con A) phytohemagglutinin P
(PHA-P), and staphylococcal enterotoxin A (SEA), to inhibit
the PFC response when added to cultures at the same time as
antigen are shown in Figure 1. Coefficients of variation for
duplicate determinations were generally less than 20%. SEA
was the most effective inhibitor, 0.01 µg/ml resulting in 88%
inhibition of the PFC response, while 0.1 µg/ml and greater con-

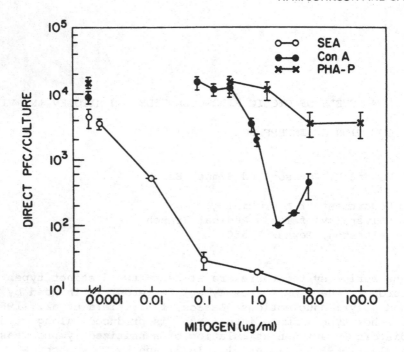

Figure 1

The suppressive effect of various T lymphocyte mitogens on the
primary in vitro PFC response to SRBC. Mitogens were added at
the time of SRBC addition, and direct anti-SRBC PFC/culture was
determined on Day 5. PFC responses are expressed as the mean of
duplicate determinations ± SD. The responses are representative
of three experiments. (From Johnson, H.M., Stanton, G.J., and
Baron, S., Proc. Soc. Exp. Biol. Med. 154:138. 1977).

centrations caused >99% inhibition of the anti-SRBC PFC response.
Con A was the second-most effective inhibitor, with 1.25 µg/ml inhibiting the PFC response by 78%. Concentrations of 2.5 µg/ml or
greater resulted in >90% inhibition of the anti-SRBC PFC response.
Interestingly, 2.5 µg/ml of Con A was more inhibitory than 5 and
10 µg of Con A. PHA-P was the least effective inhibitors of the
PFC response; 10 µg/ml was required for 76% inhibition.

     Representative data demonstrating the relative abilities
of Con A, PHA-P, and SEA to stimulate mouse spleen cell cultures
to produce immune interferon are presented in Figure 2. Coefficients
of variation for duplicate determinations were generally less than
25%. The interferon assays were carried out in cultures separate
from those used in the PFC responses, because of the differences
in incubation times of the two systems. The data in Figures 1 and

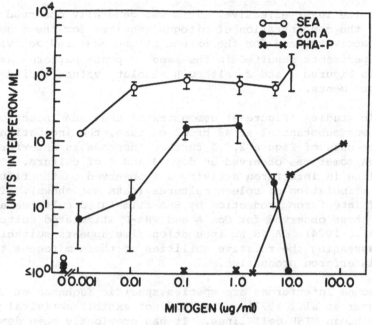

Figure 2

Stimulation of the production of immune interferon in C57Bl/6 mouse
spleen cell cultures by various T lymphocyte mitogens. Spleen cells
and mitogens were incubated for 48 hr under conditions as described
for the PFC response. Interferon concentrations are expressed as
the mean of duplicate determinations ± SD. The SD (not plotted)
for 0.001 μg of SEA is 146. The responses are representative of
three experiments. (From Johnson, H. M., Stranton, G. J., and Baron,
S. Proc. Soc. Exp. Biol. Med. 154:138. 1977).

2 are evaluated, then, for broad correlations at various mitogen
concentrations, rather than at a specific mitogen concentration.
SEA was the most effective stimulator of interferon; 134 NIH
reference units of interferon/ml were produced at 0.001 μg/ml and
644 to 1323 units/ml at 0.01 to 10 μg/ml, respectively. At 0.01 μg
of Con A/ml, only 13 units of interferon were produced. At 5 and
10 μg of Con A/ml, the amount of interferon produced significantly
declined. This was probably due to the fact that Con A was cytotoxic
at these concentrations. The Con A interferon data are consistent
with the slight recovery of the PFC response (Figure 1) at the higher
concentrations of Con A. PHA-P was the least effective inducer of
interferon; 10 μg stimulated 100 units. Lesser concentrations of
PHA-P did not stimulate the production of interferon in the cultures.
SEA was the most effective inhibitor of the PFC response and was
the best inducer of immune interferon, followed by Con A, with

PHA-P being the least effective. This was generally observed both
in terms of the concentration of mitogen required for the observed
biological activity, and for the extent of the measured activity.
Repeated experiments resulted in the same response patterns as illustrated in Figures 1 and 2, although absolute values varied
between experiments.

Kinetic studies (Figure 3) demonstrated that SEA induction of
interferon was substantial at 48 hr of culture, the incubation time
used for the data of Figure 2. A further increase in interferon
activity was sometimes observed on days 3 and 4 of culture. Occasio
ally a decline in interferon activity was observed on the fifth day
of mitogen stimulation of spleen cultures (data not shown). The
kinetics of interferon production by SEA-stimulated cultures are
similar to those observed for Con A and PHA-P stimulated cultures
(Stobo et al., 1974). A 48 hr incubation time appears suitable,
then, for assessing the relative abilities of these mitogens to
stimulate interferon production.

The immune interferons are species specific (Johnson et al.,
1976; Youngner et al., 1973). They did not exhibit antiviral
activity in human WISH cell lines. It has previously been demonstrated that the immune interferon is antigenically distinct
from virus-type interferon (Johnson et al., 1976; Youngner et al.,
1973); antibodies to virus-type interferon blocked the antiviral
activity and PFC inhibitory activity of virus-type interferon,
while having no inhibitory effect on immune interferon.

The data are consistent with the interpretation that immune
interferon might be the mediator of mitogen-induced suppressor
cell activity. Further, it is possible that immune interferon,
mitogen-induced macrophage migration inhibitory factor (MIF), and
mitogen-induced soluble immune response suppressor (SIRS) may be
biological expressions of the same substance (Johnson et al.,
1976; Youngner et al., 1973; Rich et al., 1974; Tadakuma et al.,
1976). It has not been possible to separate MIF activity (Tadakuma
et al., 1976). The interferon assay system, as demonstrated
here, may be a convenient technique for quantitating these biologica
activities.

Prerequisite to purification of immune interferon for character
ization and immunization for specific antibodies is a determination
of the most suitable inducers. We have shown here that SEA is the
most potent inducer of immune interferon in the mouse system and
that this inducer is highly active over a wide range of concentratic
The data presented here are consistent with and supportive of previous studies indicating a relationship between immune interferon
and regulation of the immune response by suppressor T cells (Johnson
et al., 1976).

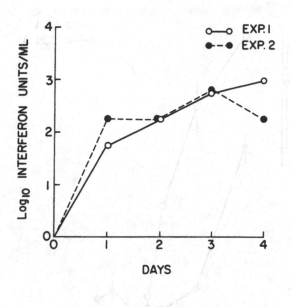

Figure 3

Kinetics of SEA stimulation of interferon production in C57B1/6
mouse spleen cell cultures. SEA was used at 0.5 µg/ml. The SD
(not shown) is similar to those of interferon assays in Fig. 2.
(From Johnson, H. M., Stanton, G. J., and Baron, S., Proc. Soc. Exp.
Biol. Med. 154:138.   1977).

Preliminary studies in humans, using peripheral blood lymphocytes, also suggest that SEA is more suitable than Con A and PHA-P
for induction of immune interferon (M. Langford, G.J. Stanton, and
H. M. Johnson, unpublished data).

It has been proposed that adenosine 3', 5'-cyclic monophosphate
(cAMP) has an inhibitory effect on the immunological and inflammatory
functions of lymphocytes (Bourne, et al., 1974). Evidence has been
obtained that suggests that cAMP may play a role in regulation of
interferon production by lymphocytes and suppression of the in vitro
PFC response (Johnson, 1977). Figure 4 presents data showing dibutyryl cAMP inhibition of the production of interferon in C57B1/6
mouse spleen cell cultures that were stimulated by the two T
cell mitogens Con A and SEA. Dibutyryl cAMP, 2 X 10^{-5}M, inhibited
Con A stimulation of interferon production by 95%, while 1 X
10^{-4}N inhibited SEA stimulation of interferon production by 85%. When
the same concentrations of dibutyryl cAMP were added to the supernatant fluids of mouse spleen cell cultures after complete interferon
production, no inhibition of antiviral activity of the interferon was
observed, so it is concluded that dibutyryl cAMP blocked the production

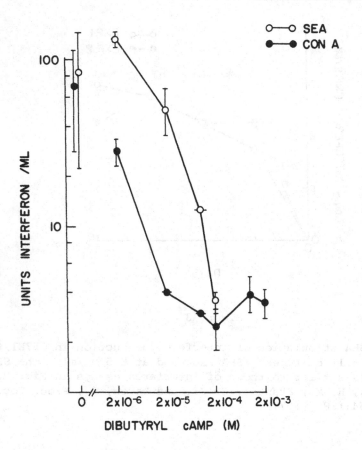

Figure 4

Dibutyryl cAMP inhibition of the production of interferon in concanavalin A (Con A) and staphylococcal enterotoxin A (SEA) stimulated
C57Bl/6 female (8 wks old) mouse spleen cell cultures. Cultures
consisted of 1.5 X 10^7 spleen cells/ml and 2 μg and 1 μg Con A and
SEA, respectively. Dibutyryl cAMP and mitogens were added to cultures at the same time and the cells were incubated for 48 hr.
Supernatant fluids were obtained by centrifugation of the harvested
cultures at 1000 RPM in a RC-3 Sorvall centrifuge at 7°C. Data
are plotted as units of interferon per ml ± SD for duplicated
samples. (From Johnson, H. M., Nature 265:154. 1977).

of mitogen-induced interferon, and not the established activity of
produced interferon.

    Table 1 presents data on dibutyryl cAMP blockade of mitogen
induced suppression of the in vitro PFC response to SRBC. SEA in-

Table 1

Effect of dibutyryl cAMP on mitogen induced suppression of the in vitro PFC response to SRBC.

Mitogen	Mitogen Conc. (ug per ml)	Dibutyryl cAMP (1.4×10^{-4}M)	PFC per 10^6 viable cells ± S.D.
SEA	0.25	-	64±8
SEA	0.50	-	101±11
SEA	0.25	+	5080±1406
SEA	0.50	+	5508±296
-	-	+	4148±296
-	-	-	5206±3626
Con A	1.0	-	7±9
Con A	2.0	-	16±11
Con A	1.0	+	1577±505
Con A	2.0	+	1456±307
-	-	+	3146±537
-	-	-	2571±691

From Johnson, H.M., 1977. Nature 265:154.

hibited the PFC response by >90% when compared to controls. Protection of the PFC response was essentially complete when dibutyryl cAMP (1.4×10^{-4}M) was added to cultures along with SEA. Dibutyryl cAMP protection of the PFC responses from Con A suppressions was also observed. Con A at 1.0 and 2.0 μg per ml inhibited the PFC response by >90% when compared to the controls. Dose response studies with various concentrations of dibutyryl cAMP showed that the concentrations (1×10^{-4} to 2×10^{-4}M) of the cyclic ribonucleotide that blocked the development of interferon in spleen cultures stimulated by the T cell nitogens (Figure 5). Dibutyryl guanosine 3', 5'-cyclic monophosphate (cGMP), at the same concentrations, had no effect on either mitogen stimulation of interferon production or mitogen induced suppression of the in vitro PFC response.

The effect of dibutyryl cAMP on SEA-induced immunosuppression and interferon production was further explored by adding dibutyryl cAMP 91.4 X 10^{-4}MO to cultures at various times relative to SEA (0.5 μg/ml) addition and determining the PFC response (Figure 6). In a parallel study, SEA induction of interferon under the same conditions was determined (Figure 6). In a parallel study, SEA induction of interferon under the same conditions was determined (Figure 6). When dibutyryl cAMP was added to the cultures at either 1 or 0 hr, complete

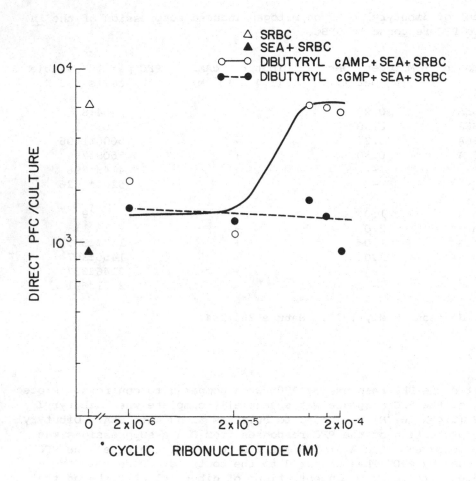

Figure 5

Dibutyryl cAMP blockade of SEA (0.5 µg/ml) suppression of the in vitro anti-SRBC PFC response in C57Bl/6 female (8 weeks old) mouse spleen cell cultures. SEA, SRBC, and cyclic ribonucleotides were added to cultures at the same time and direct anti-SRBC PFC/culture were determined on day 5. The data are expressed as the mean of duplicate determinations. Mean coefficient of variation for all determinations was 27%.

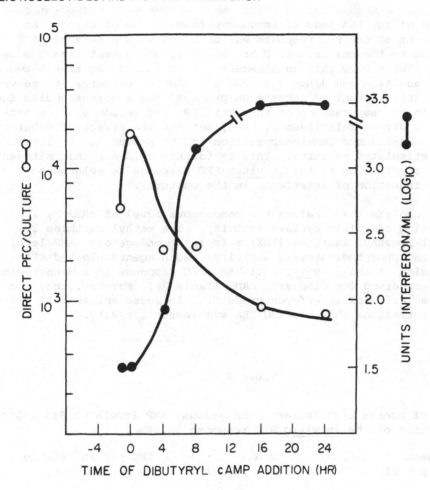

Figure 6

Determination of the PFC response and interferon production after
addition of dibutyryl cAMP to spleen cell cultures at various times
relative to SEA addition.   SEA (0.5 μg/ml) and SRBC were added at
0 hr to the spleen cells for the PFC response and dibutyryl cAMP
(1.4×10^{-4}M) was added at the indicated times relative to time 0.
Direct anti-SRBC responses were determined on day 5.   Mean PFC re-
sponses/culture ± S.D. for the SRBC control and the SEA suppressed
control were 4610±353, respectively.   Parallel studies were carried
out on interferon production under the same culture conditions,
except that SRBC were absent from the cultures and the cultures were
incubated 48 hr after SEA addition at time 0.

blockade of the SEA-induced immunosuppression was observed. An
enhancement of the PFC response was obtained when dibutyryl cAMP
was added to the cultures at 0 hr, however, subsequent experiments
did not always show this enhancement. With increasing time between
mitogen addition and dibutyryl cAMP addition to cultures, there was
an increasing amount of interferon produced and a corresponding increase in the suppression of the anti-SRBC PFC response. The data
suggest a direct relationship, then, between the effect of dibutyryl
cAMP on SEA-induced immunosuppression and on production of interferon
by SEA-stimulated cultures. This is further evidence that mitogen-
induced suppression of the in vitro PFC response is related to
mitogen induction of interferon in the cultures.

Cholera toxin raises the endogenous level of cAMP by
stimulating adenylate cyclase activity. The methyl xanthine 3-
isobutyl-1-methyl xanthine (IMX) raises the endogenous cAMP level by
inhibiting phosphodiesterase activity. Both agents blocked SEA
suppression of the in vitro anti-SRBC PFC response in a manner similar
to that observed for dibutyryl cAMP (Table 2). Further, they blocked
SEA stimulation of interferon production in mouse spleen cell cultures
at concentrations that blocked SEA suppressor activity.

Table 2

Effect of agents that increase endogenous cAMP levels on Sea induced
suppression of the in vitro PFC response to ARBC.

Agent (ug per ml)	SEA (0.5 ug per ml)	PFC per 10^6 viable cells ± S.D.
-	+	431±135
Cholera toxin (0.1)	+	2984±596
Cholera toxin (0.1)	-	4791±165
-	-	4326
-	+	91±42
IMX (12.5)	+	3973±477
IMX (12.5)	-	1366±571
-	-	2294±1238

From Johnson, H. M. 1977. Nature 265:154

Dibutyryl cAMP (10^{-3} to 10^{-7}M) did not affect the ability of
dibutyryl cAMP (1.4 X 10^{-4}M) to block the suppression of the PFC
response by SEA; nor was the effect of dibutyryl cAMP on mitogen

stimulation of interferon affected by dibutyryl cGMP. This is further
evidence that the phenomena reported here are due to cAMP, and are
not influenced by cGMP in this system.

The data suggest that mitogen induced interferon is associated
with mitogen induced suppressor cell activity, and that the interferon may possibly be a mediator of such activity. In related studies, it has been shown that high concentrations (10^{-3}M) of dibutyryl
cAMP can inhibit the yield of interferon from phytohemagglutinin-
stimulated human peripheral leukocytes (Epstein et al., 1976)
and from virus or polyribonucleotide-stimulated mouse L cell cultures
(Dianzani et al., 1972). The data presented here show a logical
relationship between cAMP, immune interferon activity, suppressor
cell activity, and regulation of the immune response. Preliminary
studies on the effect of cAMP on induction of virus-type interferon
in mouse spleen cultures suggest that this interferon system may be
slightly more resistant to blockade of induction.

Finally, comparative studies have begun on the antiviral and
in vitro immunosuppressive properties of virus-type interferon and
immune interferon (H.M. Johnson, unpublished data). Virus-type
interferon suppression of the in vitro antibody response to SRBC was
blocked by 5 X 10^{-5}M 2-mercaptoethanol (2-ME). 2-ME was capable of
blocking the suppression when added to cultures up to 48 hr after
interferon. 2-ME blockade of virus-type interferon immunosuppression
was not due to the immuno-enhancing property of 2-ME. Similar effects
of 2-ME were observed in immunosuppression by virus-type interferon
and immune interferon inducers. The data suggest that the immunosuppressive properties of these two interferons involve different
mechanisms.

2-ME did not affect the antiviral properties of either virus-
type interferon or immune interferon in nonlymphoid cells. Further,
the induction of virus-type interferon in spleen cells was neither
inhibited nor enhanced by 2-ME, while the induction of immune
interferon was enhanced. This enhancement is consistent with 2-
ME enhancement of the immunosuppressive effects of immune interferon
inducers.

There are two possibilities for 2-ME blockade of the immunosuppressive effect of virus-type interferon, while not affecting its
antiviral property. Firstly, the immunosuppressive and antiviral
properties of virus-type interferon may involve different mechanisms
at the subcellular level. Secondly, the selectivity of the blockade
by 2-ME could be due to the fact that spleen cells are the target
cells in immunosuppression, while L cells are the target cells in
inhibition of virus replication. To our knowledge the 2-ME data are
the first demonstration of an established difference in biological
properties of virus-type interferon and immune interferon.

PRODUCTION OF INTERFERON IN HUMAN CELL CULTURES BY A NEW, POTENT

VIRAL INDUCER

Patricia Jameson, Ph.D. and Sidney E. Grossberg, M.D.

Department of Microbiology
The Medical College of Wisconsin
Milwaukee, Wisconsin 53233 U.S.A.

Abstract

A newly discovered double-stranded RNA inducer of interferon, bluetongue virus (BTV), stimulates the production of large amounts of interferon in animals as well as in many types of mammalian cell cultures, including human leukocytes, and continuous cell lines. The exceptional pH lability of BTV and its lack of pathogenicity for man further recommend its use as an interferon inducer. Among several human cell lines tested, the most efficient producer of interferon was a continuous cell line designated HT-1376, derived from a bladder carcinoma. With infectious BTV as the inducer, the HT-1376 line produced more interferon per cell than did leukocytes; interferon yields ranged from 10,000 to 60,000 units per ml of crude, unconcentrated supernatant fluid. Noninfectious BTV, inactivated by ultraviolet irradiation, was as effective as infectious virions. The interferon produced in HT-1376 cells has physico-chemical and antigenic properties resembling those of fibroblast interferon produced in diploid cells.

Introduction

Recent results of limited clinical studies (1, 2, 3) and considerable experimental evidence (4, 5) attest to the potential value of interferon in the treatment of human viral diseases and cancer. Cantell and his colleagues have developed a very good method using a viral inducer to produce potent leukocyte interferon in quantities sufficient for such clinical trials (6, 7, 8, 9). Because of the encouraging results obtained with that interferon, it appears that the needs for larger quantities of interferon may

37

be growing faster than the means of supplying it. There is thus
increasing need for convenient, efficient, and economical methods
to produce large quantities of interferon of consistent potency,
stability, and purity.

Strander et al. have stated that the production of leukocyte
interferon "... will always be limited by the availability of
human blood" (10). The demand for leukocyte interferon may be
moderated to some extent by a preference for interferon of fibroblast origin for some applications in cancer therapy. Such needs
may develop, for example, from the recent observations that
various types of tumor cells may respond better to the anticellular
action of the fibroblast interferon species (11). Available
methods for the production of potent fibroblast interferon employ
the superinduction process (12, 13, 14, 15), which may require
minor adjustments, such as modification of actinomycin D concentration, for each cell strain (J. Vilcek, personal communication,
1977) and may pose problems for use on a large scale. Since the
improvements being made in interferon purification methods should
permit its separation from inducers and cell culture by-products,
all available means of production should be considered.

To obtain a reproducible system for the production of large
quantities of interferon, we have searched for (i) a continuous
cell line capable of producing high yields of interferon, and (ii)
a viral inducer that would be highly effective in human cells. The
cell lines tested for interferon production were selected on the
basis of cell type, availability, and growth potential. The pro-
spective value of any given cell was estimated by its response to
at least two different inducers, and interferon yields were compared
with those obtainable with lymphocyte cultures. Three inducers
were selected on the basis of their proven effectiveness in other
cells (7, 16): polyriboinosinate·polyribocytidylate complexed with
DEAE-dextran, Sendai virus, and Newcastle disease virus (NDV). A
fourth, bluetongue virus (BTV), a member of the orbivirus subgroup
of the double-stranded RNA-containing reovirus family, is the most
potent interferon inducer in mice we have found, consistently
producing more that 100,000 units/ml in plasma after injection of
10^8 PFU/mouse (17,18). BTV was consistently an excellent inducer
in rabbit kidney cell cultures, producing an average of 40,000
units/ml in more than 20 different experiments; these yields made
it possible to produce the high-titered reference interferon standard G019-902-528 for the National Institutes of Health. BTV has two
other properties which add to its suitability as a desirable interferon inducer: (i) its extreme pH lability (19) permits total
inactivation of residual virus within one hour of exposure to pH 2,
and (ii) the virus has no known human pathogenicity (20); D. W.
Verwoerd, personal communication, 1976) a safety factor to be
considered in large-scale production.

Material and Methods

Cell cultures. Fresh lymphocyte suspensions were prepared according to the methods of Cantell et al. (7) except that the EDTA concentration was 0.45%, and the cells were not routinely pooled during storage at 4°C or 20°C. The buffy coats were provided by the Milwaukee Blood Center (Dr. R. Duquesnoy) from donations used for platelet-phoresis. Human cell lines were obtained from several investigators and propagated according to published methods. Dr. M. Ahmed, Pfizer Research Laboratories, provided the NC-37, HEK (21), and KC (22) cell lines. Dr. S. A. Aaronson, National Institutes of Health (NIH), gave us the A375 melanoma and A549 lung carcinoma cell lines (23). Dr. S. Rasheed, University of Southern California School of Medicine, Los Angeles, made available the HT-1417 lymphoblastoid cell line and the HT-1376 bladder carcinoma cell line (24) which we propagated in RPMI-1640 medium supplemented with 10% fetal bovine serum (FBS) for HT-1376 cells, and 20% FBS for HT-1417 cells. NB cells (25), obtained from Dr. M. Oxman, University of California, San Diego, and HuF/A28 cells (26), from Dr. J. Butel, Baylor University College of Medicine, Texas, are SV40-transformed cell lines. The lymphoblastoid cell line 8392B (27) was received from Dr. M. Essex, Harvard University School of Public Health, and was propagated in medium RPMI-1640 supplemented with 20% FBS. HEK, KC, and HuG/A28 cell lines were grown in Minimal Essential Medium with Earle's salts (EMEM), supplemented with 10% FBS; NB, A549, and A375 cells were propagated in Dulbecco's modified MEM with glucose increased to 4.5mg/ml and containing 10% FBS.

Viruses. Bluetongue virus (BTV) serotype 10 (28) was purchased from Colorado Serum Co. (Denver, Co.) as a vaccine strain designated BT-8. Virus was propagated as described by Verwoerd (29) with the following modifications: the inoculum multiplicity was 5 to 10 pfu/cell, the virus attachment period was one hour, and the culture medium contained 5% calf serum. BTV was titrated by plaque assay in L cell monolayers (30); DEAE-dextran was added to the agar overlay (400 µg/ml). Sendai virus (NIH V321-011-000) was propagated in chicken embryos and titrated in primary hamster kidney cell monolayers (31). Newcastle disease virus (NDV), CG strain, was prepared and titrated as described previously (32).

Interferon induction. Interferon was induced in leukocyte cultures by the method of Cantell (7) except that cells from different donors were not routinely pooled and cultures were maintained in Erlenmeyer flasks or roller bottles, priming with interferon was omitted, and the medium used during interferon induction contained 0.2% human plasma protein fraction (HPPF) instead of 4% human serum and had the usual phosphate and bicarbonate concentration for MEM as well as added tricine buffer.

Induction of interferon in lymphoblastoid cell lines was the

same as the standard procedure for leukocyte cultures. Other cells
were induced by the attachment of virus to washed confluent
monolayers of cells grown for approximately 4 to 6 days in tubes,
30 ml plastic flasks, or 32-oz. prescription bottles. At the end
of the attachment period, EMEM containing non-essential amino acids
and only 2% FBS was added at half the volume used for cell growth.
Supernatant fluids were collected 24 hours after the virus inoculation and stored at $-70^{\circ}C$. Residual inducer virus was inactivated
by pH 2 treatment before titration of interferon: by adjustment to
pH 2 for 1 hour with BTV or by dialysis against pH 2 buffer at $4^{\circ}C$
for 1 day with Sendai virus, or for 5 days with NDV. We have confirmed the inordinate lability of BTV to low pH. Crude virus suspensions (made up in medium with 5% serum) lost 10^5 pfu in 30 seconds
and $>10^7$ pfu in 1 minute at pH 2.

For induction of interferon by polyriboinosinate·polyribocytidy-
late--poly(I)·poly(C)--(P-L Biochemicals, Milwaukee, WI.), medium
was decanted, a small volume of the poly(I)·poly(C):DEAE-dextran
mixture, at the concentrations stated for each experiment, was
incubated with cells for 1 hour at $37^{\circ}C$, and medium was added
without removing the inoculum. In some experiments cells were
treated with poly(I)·poly(C) alone, then supplied with medium containing neomycin (33).

Interferon assay. Interferon was quantitated by a hemagglutinin
yield-reduction assay using encephalomyocarditis virus (34) with the
BUD-8 cell strain (35). This assay detects approximately 4- to 6-
fold more activity that the assigned titer for the NIH reference
leukocyte interferon (G023-901-527).

Interferon characterization. Trypsin sensitivity was tested
by incubation with crystalline trypsin (Worthington Biochemical Corp.
Freehold, N.J.), 2.5 mg/ml for 1 hour at $37^{\circ}C$. The reaction was
stopped by the addition of 0.3 mg of soybean trypsin inhibitor,
and the antiviral activity was titrated in the treated sample
and appropriate reagent controls. Heat stability was estimated by
incubation of replicate samples at $56^{\circ}C$ for 30, 60, and 120 minutes.
Acid resistance and dialyzability were established during dialysis
for 1 day against pH 2 HCL-KCl buffer, 0.15 M. Neutralization by
specific antisera provided by the NIH was determined by titration
of remaining antiviral activity after incubation of mixtures of
serum dilutions with a constant amount of interferon for 30 minutes
at $37^{\circ}C$. Heterologous activity was assessed in interferon assays
using L cells (36) and RK-13 cells (34).

Leukocyte interferon. Crude leukocyte interferon was provided
by K. Cantell, Central Public Health Laboratory and Finnish Red Cross
Blood Transfusion Service, Helsinki, Finland.

Results

Interferon induction in leukocytes. The interferon-inducing effectiveness of BTV was compared with that of Sendai virus in leukocyte cultures (Table 1). The yield obtained with BTV was at least 10 times that induced by Sandai virus with input viral multiplicities of 1.0 and 0.5 PFU/cell, respectively. Similar results were obtained whether or not the maintenance medium contained phosphate buffer. Priming the cells with interferon, 100 units/ml for 2 hours before the addition of inducer, did not enhance yields as it did for others(9). Interferon production was not altered if virus was added directly to the final cell suspension or adsorbed to concentrated cells in a small volume. Whereas input multiplicity (m.o.i.) affected induction with Sendai significantly (m.o.i. >38 gave yields of 33,000 units/ml, BTV induced yields of 33,000-60,000 units/ml over a range of m.o.i. of 0.25 to 5 (Table 2).

Since De Maeyer et al. (37) had shown that genetic differences in mice have a quantitative effect on circulating interferon produced by hemopoietic cells with a given inducer, interferon induction by BTV was used to test the response of leukocytes from individual donors of different blood group phenotypes. Results are tabulated for cell responses associated with each of the four ABO groups (Table 3). There did not appear to be any correlation between blood group and interferon yield. Leukocytes from an individual within any given blood group varied considerable in their interferon responses. No correlations can yet be made between variations in response and HLA type (data not presented). Several minor modifications of procedure did not markedly improve interferon yields, including changes in cell storage temperatures, alternative agitation methods, UV inactivation of BTV, reduction of medium volume, or substitution of a combination of tripeptide, oleate and bovine serum albumin for serum in the culture medium (data not presented).

Satisfactory interferon yields could be obtained from 500 ml cultures in 750 cm^2 roller bottles turning at 15 revolutions per hour as well as from 200 ml cultures in Erlenmeyer flasks shaken at 120 revolutions per minute.

Interferon induction in cell lines. Interferon induction by BTV, NDV, Sendai virus, and poly(I)·poly(C) was compared in a number of cell lines derived from tumors or normal tissues (Jameson and Grossberg, manuscript in preparation). All of the cell lines, HuF/A28, failed to respond to any of the viruses; all of the seven tested with BTV made some interferon. Most cells responded as well, or better, to BTV as to the other inducers. The cell lines could be ranked into three groups based on their ability to produce interferon with the four inducers. HT-1376, NC37, A549, and NB cell lines were good producers; HEK, A375, HT-1417, 8392B, and KC cell lines, marginal producers; HuF/A28 was clearly a poor producer since it failed to

Table 1

Induction of Human Interferon in Leukocytes

Induction conditions[1]	Titer of interferon obtained with the indicated medium	
	MEM with phosphate	MEM without phosphate
Sendai virus with priming[2]	375	375
Bluetongue virus with priming	50,000	55,000
Bluetongue virus	64,000	64,000

1. Viruses were added to a suspension of 10^7 cells/ml. The input multiplicity for BTV was 1 PFU/cell and for Sendai virus was 0.5 PFU/cell.

2. Priming was done for 2 hours with leukocyte interferon (100 units/ml of leukocyte interferon) before addition of inducing virus.

Table 2

Interferon Yield from Leukocytes in Relation to
Virus Input Multiplicity

Virus	Multiplicity [1]	Interferon yield (units/ml)
Sendai	3.8	2,100
	7.6	6,400
	15.2	6,400
	38	33,000
	76	33,000
Bluetongue	0.25	33,000
	0.5	33,000
	1.0	33,000
	2.5	60,000
	5.0	60,000

1. Inoculum multiplicity was varied by the use of different
   columes of undiluted stock virus, 0.05 to 1.0 ml, to infect
   the suspension culture. The multiplicities are based on
   plaque titration of BTV in L cells, and Sendai virus in
   primary hamster kidney cells.

Table 3

Interferon Induction by BTV in Leukocytes of Donors Having Different Blood Group Phenotypes[1/]

Interferon yields[2/](units/ml) in the stated number of leukocyte cultures from donors of the indicated blood group

Parameter	O	A	B	AB
Geometric mean	5,120[3/]	6,840	5,310	9,560
n	6	5	4	6
Range:				
low	<200[3/]	1,300	1,200	540
high	64,000	17,500	34,000	27,000

1. Cells were held overnight at 4°C or 20°C in buffer containing 0.4% EDTA and separated the following day by the ammonium chloride method described by K. Cantell (7). Interferon was induced by adding BTV at a multiplicity of 1 PFU/cell to lymphocyte suspensions of 10^7 cells/ml in MEM supplemented with 0.2% HPPF. Interferon was collected at 24 hours and cells were removed by centrifugation; residual BTV was inactivated by pH 2 treatment at 4°C for 1 hour. All but 2 or the donors were Rh positive.

2. Geometric mean titers were calculated because of the extreme variations in yields as indicated by the range.

3. Two yields below the limits of the dilution range tested (<200) were excluded from the calculation; n = 6 for G.M.T., n = 8 for range.

respond to the viral inducers and produced only 15 units/ml with
poly(I)·poly(C). The best interferon yields were obtained with a
bladder carcinoma cell line, HT-1376, which responded to all
four inducers.

Interferon production by HT-1376 cells. Varying the BTV
multiplicity from 0.1 to 10.0 PFR/ml did not affect the level of
interferon produced by HT-1376 cells. Better yields of interferon
were obtained with larger cultures (Table 4). HT-1376 cultures
exposes to BTV, 10 PFU/cell, in 32 oz. bottles produced yields
of 10,000 to 60,000 (arithmetic mean of 21,000 units/ml or geometric mean titer 18,600 units/ml in 22 experiments over a period
of 1 year). These titers are comparable to the best yields we
obtained with leukocyte suspensions. However, the yields with HT-
1376 cells were much more consistently obtained than with leukocyte
cultures in which titers ranged from <200 to 64,000 units/ml
(geometric mean titer of 6,600 units/ml in leukocyte cultures containing 10^7 cells/ml induced with BTV, 1 PFU/cell, the same method
employed in 23 experiments over a 7 month period). The interferon
yield (per 10^3 cells) from HT-1376 cells was 19 units compared to
0.66 units from leukocytes.

Since vessel size unaccountable can influence interferon yields,
the relative efficiency of three viral inducers was compared with
poly(I)·poly(C) in 32 oz bottle cultures of HT-1376 cells; the
optimum dose of each inducer was used (Table 5). BTV induced the
most interferon; NDV and Sendai virus were used at multiplicities
10 times those of BTV. Even when totally inactivated by UV irradiation, BTV was an effective interferon inducer in HT-1376 cells
as well as in leukocytes.

Little improvement in yield of interferon was obtained either
by increasing the concentration of poly(I)·poly(C), by adding
DEAE-dextran to the polynucleotide, or by adding neomycin to the
culture medium (Table 6). BTV is therefore about 5 to 6 times more
effective an interferon inducer in HT-1376 cells than is poly(I)·
poly(C).

Properties of the antiviral substances induced by BTV in
leukocytes and HT-1376 cells. The physicochemical properties of the
antiviral substances produced by leukocytes and HT-1376 cells induced
with BTV conformed to those described for leukocyte and fibroblast interferons, respectively. Both were stable at pH 2, non-
dialyzable, and destroyed by trypsin. The substance produced by
the HT-1376 cells resembles that of fibroblast interferon with respect to: its relative heat lability (6% of original activity
remained after 2 hours at 56°C (38); its neutralization by antibody
to fibroblast interferon serum (39); and its molecular weight
(determined by chromatography on Sephadex G-200) of 26,000 daltons

Table 4

Interferon Induction by BIV in HT-1376 Bladder Carcinoma Cells

Conditions of induction [1]			Interferon yield	
Number of cells	Vessel [2]	m.o.i. of BTV	units per ml	units per cell x 10^3
5.0×10^5	tubes	50	4,000	8
5.0×10^5	tubes	50	2,600	5
5.0×10^5	tubes	50	2,000	4
5.0×10^5	tubes	40	16,500	33
2.5×10^6	flasks	10	62,000	62
1.2×10^7	bottles	10	21,000 [3]	26 [3]

1. Virus was absorbed for 1 hour in a volume of 0.25 ml/tube, or flask, and 1 ml/ bottle. Inducer was then removed and cells were washed before the addition of medium (1 ml/tube, 2.5 ml/flask, or 15 ml/bottle).

2. Disposable borosilicate test tubes were 16 x 125 mm, plastic flasks were 25 cm^2, and flint-glass prescription bottles were 32 oz capacity with a surface area of 120 cm^2.

3. This number represents 22 inductions with BTV giving an average interferon yield of (units/ml) 21,000 with a nimimum of 10,000 and a maximum of 62,000.

Table 5

Interferon Induction in HT-1376 Cells by Four Different Inducers

Titer (units/ml) of interferon induced in replicate cultures using the indicated dose of inducer[1]

BTV (m.o.i. = 3)	NDV (m.o.i. = 50)	Sendai virus (m.o.i. = 30)	poly(I)·poly(C) (100 µg/ml)
10,500	5,000	1,100	1,100
7,500	5,000	1,100	1,100

1. Two cultures of HT-1376 cells in 32 oz. prescription bottles (about 1.5×10^6 cells corresponding to 10^5 cells/ml) were treated with the inducer indicated. Virus inoculum multiplicities (m.o.i.) are given as PFU/ml.

in its disaggregated form at pH 2, or predominantly 2000,000 daltons
at pH 7 (J. Sedmak and S. E. Grossberg, unpublished). BTV induced
leukocyte antiviral activity is relatively heat stable (63% remaining
after 2 hours at 56°C) and is neutralized by antibody to leukocyte
interferon.

Table 6

Interferon Induction in HT-1376 Cells Using poly(I)·poly(C)

Adjuvant used	Concentration of adjuvant in culture medium (µg/ml)	Interferon units/ml produced by cells induced with poly(I)·poly(C) at the indicated concentration	
		10 µg/ml	100 µg/ml
None		950; 1,600[1]	2,900
DEAE-dextran	1	950; 930	nd[2]
	10	<100; 235	2,700
	100	nd; 1,600	2,900
Neomycin	300	1,150; nd	4,500

1. The results of two experiments are shown
2. Not done

Discussion

Bluetongue virus was tested in human cell cultures because it
is the most potent interferon inducer we have ever tested in mice;
it is capable of stimulating yields of up to 500,000 units/ml in
plasma (17). In cell culture BTV is equal to or better than NDV
and Sendai virus. Two other viral properties make BTV the inducer
of choice in preference to NDV or Sendai virus: (i) it is extremely
pH labile (19) and thus may be completely inactivated by a safe
margin within 1 hour at pH 2; and (ii) the virus has no known human
pathogenicity (20); D. W. Verwoerd, personal communication, 1976),
which makes it safe to handle in large quantities.

The exact mechanism of interferon induction by BTV has not been
established. Extensive studies with the relatively well-character-
ized mutants of the related reoviruses have not revealed the precise

nature of the inducing principle provided by infection of cells with
intact virus (40).   It is clear that isolated viral double-stranded
RNA is an effective inducer (41).   The role of defective particles
in the interferon induction process has not been determined for
either BTV or the paramyxoviruses, NDV and Sendai.

The HT-1376 cell line was developed and characterized extensive-
ly by Rasheed et al., (23), and is available from the American Type
Culture Collection. The cell is of epithelial origin, having a
glucose-6-phosphate dehydrogenase isozyme mobility of type B, a
modal chromosome number of 104-121, and fibrinolytic activity. No
virus particles or mycoplasma were seen by electron microscopy,
no reverse transcriptase activity was detected, and no reaction was
obtained with antisera prepared against p30 antigen of C-type viruses
were induced by chemical treatment . No antigens of type 2
herpesvirus have been detected (B. Molholt, personal communication,
1977). The antigenic, morphological and biological diversity among
several different cell lines established from bladder carcinomas,
including HT-1376, suggest that they do not share a common viral
etiology. Growth characteristics include a doubling time of 50
hours, with a saturation density of 1.6×10^7 cells/cm^2, and a
plating efficiency of 36% (24).

Interferon production by the HT-1376 cell line, yielding 19
units/10^3 cells. Additional modifications of production conditions
may further improve yields. Nevertheless, the demonstrated interferon producing ability of the HT-1376 cell line approaches that of
commonly used interferon-producing cell strains, namely 20 to 50
units/10^3 cells (12,13,14), and the HT-1376 cell line can be propagated indefinitely with no limit to cell number. There is, therefore, considerable potential value of this, or other, continuous
cell lines in providing the interferon needed for clinical testing
and application. In addition, the HT-1376 cell line may further
offer a suitable in vitro screening system for the evaluation of
interferon inducers for use in man.

Acknowledgment

We are grateful to Mmes. Mary Dixon and Christine K. Schoenherr
for excellent technical assistance, to Dr. B. Molholt for testing
for herpesvirus antigens, and to Dr. J. J. Sedmak for doing the
column chromatography. We thank the NIH for antisera to fibroblast
and leukocyte interferons, Dr. K. Cantell for leukocyte interferon,
Dr. R. J. Duquesnoy at the Milwaukee Blood Center for providing buffy
coats; and Drs. S. A. Aaronson, M. Ahemen, J. Butel, M. Essex,
M. Oxman, and S. Rasheed for their generous gifts of cells.

References

1. Edy, V.G., A. Billiau and P. De Somer. 1977. Large-scale
    production of interferon in human diploid fibroblast cells.
    Production of Human Interferon and Investigation of Its Clin-
    ical Use. Eds. W. Stinebring and P. J. Chapple. Proceedings
    of a Symposium Workshop of the Tissue Culture Association, this
    symposium issue.

2. Merigan, T. C. 1977. Pharmacokinetics and side effects of inter
    feron system: a current review. Eds. S. Baron and F. Dianzani.
    Texas Rep. Biol. Med. (in press).

3. Strander, H. 1977. Interferon therapy in human neoplastic
    diseases. Production of Human Interferon and Investigation
    of Its Clinical Use. Eds. W. Stinebring and P. J. Chapple.
    Proceedings of a Symposium Workshop of the Tissue Culture
    Association, this symposium issue.

4. Baron, S., F. Dianzani (Eds.). 1977. The interferon system:
    A current review. Texas Rep. Biol. Med. (in press).

5. Finter, N. B. (Ed). 1973. Interferon and interferon inducers.
    Amsterdam, North-Holland.

6. Cantell, K. 1970. Preparation of human leukocyte interferon.
    International Symposium on Standardization of Interferon and
    Interferon Inducers. Eds. F..T. Perkins, R. H. Regamey. Symp.
    Ser. Immunobiol. Standard Vol. 14:6-8. Basel, Karger.

7. Cantell, K., S. Hirvonen, K. E. Mogensen et al. 1974. Human
    leukocyte interferon: Production, purification, stability, and
    animal experiments. The Production and Use of Interferon for
    the Treatment and Prevention of Human Virus Infections. Eds.
    C. Waymouth. Proceedings of a Tissue Culture Association
    Workshop, Tissue Culture Association, Rockville, MD. pp 35-38.

8. Kauppinen, H., G. Myllylä, K. Cantell. 1977. Large-scale
    production and properties of human leukocyte interferon used
    in clinical trails. Production of Human Interferon and Inves-
    tigation of Its Clinical Use. Eds. W. Stinebring and P. J.
    Chapple. Proceedings of a Symposium Workshop of the Tissue
    Culture Association, this symposium issue.

9. Tovell, D. and K. Cantell. 1971. Kinetics of interferon
    production in human leukocyte suspensions. J. Gen. Virol.
    13:485-489.

10. Strander, H., K. E. Mogenser and K. Cantell. 1975. Production
     of human lymphoblastoid interferon. J. Clin. Microbiol. 1:
     116-117.

11. Einhorn, S., H. Strander. 1977. Is interferon tissue specific?
     Effect of human leukocyte and fibroblast interferons on the
     growth of lymphoblastoid and osteosarcoma cell lines. J. Gen.
     Virol. 35:573-577.

12. Havell, E.A., J. Vilcek. 1972. Production of high-titered
     interferon in cultures of human diploid cells. Antimicrobial
     Agents Chemother. 2:476-484.

13. Havell, E.A., J. Vilcek. 1974. Mass production and some
     characteristics of human interferon from diploid cells. The
     Production and Use of Interferon for the Treatment and Prevent-
     ion of Human Virus Infections. Ed. C. Waymouth. Proceedings
     of a Tissue Culture Association Workshop. Tissue Culture
     Association, Rockville, MD p 47.

14. Havell, E. A., J. Vilcek, M. L. Gradoville, et al. 1977.
     Selection of new human foreskin fibroblast cell strains for
     interferon production. Production of Human Interferon and
     Investigation of Its Clinical Use. Eds. W. Stinebring and
     P. J. Chapple. Proceedings of a Symposium Workshop of the
     Tissue Culture Association, this symposium issue.

15. Tan, Y. H., J. A. Armstrong, Y. H. Ke, et al. 1970. Regula-
     tion of cellular interferon production: enhancement by antimeta-
     bolites. Proc. Natl. Acad. Sci. USA 67:464-471.

16. Ho, M. 1973. Factors influencing interferon production. In-
     feron and Interferon Inducers. Ed. N. B. Finter. Amsterdam,
     North-Holland. pp 73-105.

17. Jameson, P., C. K. Schoenherr, S. E. Grossberg. 1978.
     Bluetongue virus, and exceptionally potent interferon inducer
     in mice. Infect. Immun. 20:321-323.

18. Jameson, P. 1977. Induction of the interferon protein in
     vivo: virus inducers. The Interferon System: A Current
     Review. Eds. S. Baron, F. Dianzani. Texas Rep. Biol. Med.
     in press.

19. Svehag, S. E., L. Leendertsen, J. R. Gorham. 1966. Sensitivity
     of bluetongue virus to lipid solvents, trypsin and pH changes
     and its serological relationship to arboviruses. J. Hyg.
     Camb. 64:339-346.

20. Joklik, W.K. 1974. Reproduction of reoviridae. Comprehensive
    Virology. Eds. H. Fraenkel-Conrat and R. R. Wagner. New York
    Plenum Press. pp 231-334.

21. Ahmed, M., W. Korol, J. Yeh, et al. 1974. Detection of Mason-
    Pfizer virus infection with human KC cells carrying Rous virus
    genome. J. Natl. Cancer Inst 53:383-387.

22. MacIntyre, E.H., R.A. Grimes, A. E. Vatter. 1969. Cytology and
    growth characteristics of human astrocytes transformed by
    Rous sarcoma virus. J. Cell Sci. 5:583-602.

23. Giard, D. J., S. A. Aaronson, G. J. Todaro, et al. 1973.
    In vitro cultivation of human tumors: establishment of cell
    lines derived from a series of solid tumors. J. Natl.
    Cancer Inst. 51:1417-1423.

24. Rasheed, S., M. B. Gardner, W. W. Rongey, et al. 1977.
    Human bladder carcinoma: Characterization of two new tumor
    cell lines and search for tumor virus. J. Natl Cancer Inst.
    58:881-890.

25. Shein, H.M., J. F. Enders. 1962. Transformation induced by
    simian virus 40 in human renal cell cultures, I. Morphology
    and growth characteristics. Proc Natl. Acad. Sci. USA 48:
    1164-1172.

26. Brugge, J. S., J. S. Butel. 1975. Role of simian virus 40
    gene A function in maintenance of transformation. J. Virol.
    15:619-635.

27. Huang, C. C., Y. Hou, L. D. Woods, et al. 1974. Cytogenetic
    study of human lymphoid T-cell lines derived from lymphocytic
    leukemia. J. Natl. Cancer Inst. 53:655-660.

28. Howell, P. G., D. W. Verwoerd. 1971. Bluetongue virus. Virol.
    Monogr. 9:37-74.

29. Verwoerd, D. W. 1969. Purification and characterization of
    bluetongue virus. Virology. 38:203-212.

30. Howell, P. G., D. W. Verwoerd, R. A. Oellermann. 1967.
    Plaque formation by bluetongue virus. Onderstepoort. J.
    Vet. Res. 34:317-332.

31. Grossberg, S. W. 1964. Human influenza A viruses: rapid
    plaque assay in hamster kidney cells. Science 144:1246-1247.

32. Morahan, P. S., S. E. Grossberg. 1970. Age-related cellular resistances of the chicken embryo to viral infections. I. Interferon and natural resistance to myxoviruses and vesicular stomatitis virus. J. Inf. Dis. 121:615-623.

33. Billiau, A., C. E. Buckler, F. Dianzani, et al. 1969. Induction of the interferon mechanism by single-stranded RNA: potentiation by poly-basic substances. Proc. Soc. Exp. Biol. Med. 132:790-796.

34. Jameson, P., M. A. Dixon, S. E. Grossberg. 1977. A sensitive interferon assay for many species of cells: encephalomyocarditis virus hemagglutinin yield reduction. Proc. Soc. Exp. Biol. Med. 155:173-178.

35. Sedmak, J. J., S. E. Grossberg, P. Jameson. 1975. The neuraminidase yield-reduction bioassay of human and other interferons. Proc. Soc. Exp. Biol. Med. 149:433-438.

36. Oie, H. K., C. E. Buckler, C. P. Uhlendorf, et al. 1972. Improved assays for a variety of interferons. Proc. Soc. Exp. Biol. Med. 140:1178-1181.

37. De Maeyer, E., P. Jullien, J. De Maeyer-Guignard, et al. 1975. Effect of mouse genotype on interferon production. II. Distribution of If-1 alleles among inbred strains and transfer of phenotype by grafting bone marrow cells. Immunogenetics 2:151-160.

38. Vilcek, J., E. A. Havell. 1973. Stabilization of interferon messenger RNA activity by treatment of cells with metabolic inhibitors and lowering of the incubation temperature. Proc. Natl. Acad. Sci. USA 70:3909-3913.

39. Havell, E. A., B. Berman, C. A. Ogburn, et al. 1975. Two antigenically distinct species of human interferon. Proc. Natl. Acad. Sci. USA 72:2185-2187.

40. Lai, M. T., W. Joklik. 1973. The induction of interferon by temperature-sensitive mutants of reovirus, UT-irradiated reovirus, and subviral reovirus particles. Virology 51:191-204.

41. Tytell, A. A., G. P. Lampson, A. K. Field, et al. 1967. Inducers of interferon and host resistance, III. double-stranded RNA from reovirus type 3 virions (Reo 3-RNA). Proc. Natl. Acad. Sci. USA 58:1719-1722.

HUMAN INTERFERON: LARGE SCALE PRODUCTION IN EMBRYO FIBROBLAST CULTURES

V.G. Edy, J. VanDamme, A. Billiau and P. DeSomer

Department of Human Biology, Division of Microbiology,
Rega Institute, University of Leuven, B-3000 Leuven,
Belgium

INTRODUCTION

Initial exploratory tests, and trials of the use of exogenous interferon in man have had fairly promising results (1-7); however, further more extensive trials will probably be restricted by limitations in the supply of interferon. Almost all trials of interferon in man have, up to now, been performed using leukocyte-derived interferon produced by induction with Sendai virus (8). This system of production is dependent on the supply of human blood buffy coats, and this factor alone restricts expansion of production from present levels.

To supplement the supply of leukocyte interferon, two alternative sources of human interferon are currently being exploited: lymphoblast and fibroblast cultures. Initially, the only lymphoblast cultures available were poor producers of interferon, but recently a high-yielding line has been described (9), resulting in a renewal of interest in lymphoblast-derived interferon.

Until this high producer line, called Namalwa, was described, the main hopes for large-scale human interferon production lay in the use of human fibroblast cultures.

To examine the feasibility of production of relatively large amounts of human fibroblast-derived interferon, a roller bottle system has been developed using human embryo fibroblasts. Interferon induction is by treatment with the synthetic double-stranded RNA poly(I) · poly(C), and the cultures are chemically

superinduced (10-12). The cell cultures are split 1 : 2 or
1 : 4 by trypsinisation each week, and can be used for interferon
production at any passage level before senescence; the
population doubling level does not seem to affect interferon
production. Cultures are routinely screened for mycoplasma
contamination, and are tested for normal karyotype. No deviations
from the normal karyotype have been observed.

Roller bottles which are to be induced to make interferon
are kept for 10-13 days after establishment, (i.e. 8-11 days
post-confluency) and are refed once during this holding phase.
The necessity of this aging period for maximal interferon
production upon superinduction is well known (10,11), but it
has not yet been fully explained.

To reduce the possibility of contamination of the final
product with any bovine proteins, the cultures are pretreated,
before induction, with a medium in which the calf serum supplement
(routinely used for cell growth) is replaced by 4.5 mg.ml^{-1} of
human plasma protein fraction (obtained from the Belgian Red
Cross) (13). This preincubation step is also used for the
priming (14) of the cultures by including 100 U.ml^{-1} of interferon
in the medium.

After 16 hr incubation at 37°, the priming-human plasma
protein fraction medium is discarded, and replaced by serum-free
medium containing the interferon inducer (poly(I)·poly(C) at
50 µg.ml^{-1}, and 10 µg.ml^{-1} of cycloheximide. This mixture is
left on the cultures for 6 hr at 37°; actinomycin D being added
at 4 hr to a final concentration of 1 µg.ml^{-1}. This is essentially
the superinduction scheme of Havell and Vilcek (10); a similar
schedule by Billiau and his coworkers (11) gives similar yields
of interferon, but involves an extra manipulation of the cultures.
On the other hand, this latter scheme allows the re-use of the
poly(I)·poly(C) interferon inducer.

Following the induction and superinduction period, the cultures
are drained, washed several times with medium, and are then refed
with medium containing 0.45 mg.ml^{-1} human plasma protein fraction.
If all protein is omitted from this final interferon production
medium, interferon yields are very variable, and on average the
production is rather poor.

After incubation at 37° for 18 hr, the culture fluid, now
containing interferon, is harvested and centrifuged (2000 x g)
to remove any cellular debris. As the actinomycin D block on RNA
synthesis is irreversible, the cell cultures are no longer viable,
and are discarded.

Using this production system with about 100 roller bottles. week^{-1}, about $10^{8.0}$ units of crude interferon are produced. Table 1 summarises the most important points in the production system.

TABLE 1

Summary of Human Fibroblast Interferon Production

1) Cells used: Human embryo skin-muscle fibroblasts

2) Induction system: Roller cultures, 13 days old. Primed, induced with poly(I).poly (C), superinduced with cycloheximide and actinomycin D

3) Final production: EMEM + 1% (v/v) human plasma protein fraction, 20 ml per bottle.

4) Average yield: 0.5 x 10^5 U per ml
10^6 U per bottle
30 U per 10^3 cells.

5) Weekly production: 2.7 litres (1.3 x 10^8 U)

6) Crude specific Activity: 1.05 x 10^5 U per mg protein

The roller bottle culture is excellent for this scale of production, but is probably not economically suitable for use on a larger scale. Therefore various alternative schemes for large-scale monolayer cell culture have been tested. Most successful has been the "Sterilin" Bulk Culture Vessel (15, Sterilin Ltd., Richmond, Surrey, Gr. Britain), which is simple, inexpensive and easy to manipulate. Yields of interferon per cell and per square centimetre of culture surface are very similar to those achieved with roller bottles.

The choice of cell strain for interferon production does not seem to be a particularly critical point. Most embryo skin-muscle fibroblasts tested in our system yield between 25 and 50 reference units of interferon per 1000 cells.

The crude interferon produced routinely in roller cultures has a titre of between 30,000 and 64,000 reference units.ml^{-1}; although yields outside this range are not uncommon. Before it can be used for injection, the interferon has to be concentrated, and partially purified. There are very many ways of achieving this; the full range of protein chemistry has been brought to bear on interferon purification. Our own preference is for glass adsorption chromatography (16), a simple and rapid process which can be operated in a batch system. Using this process, interferon with a specific activity of 1-5 x 10^6 U.mg^{-1} protein can be readily produced.

Recently the first results of administration of human fibro-blast interferon in man have been reported (3), and several more trials are in progress.

It should be borne in mind that whilst lymphoblast and leukocyte derived interferons are quite similar in many of their physicochemical (9, Fantes, Burman and Johnston, pers. comm.) and biological (Finter, Christofinis, Whitaker and Johnston, pers. comm.) properties; fibroblast interferon is quite distinct (17-20). It is therefore possible that these two classes of interferons will prove to act with varying effectiveness (in different conditions, and that one type will not supplant the other in clinical trials or treatment.

REFERENCES

1. Ahstrom, L., A. Dohlwitz, H. Strander, and K. Cantell. 1974.
   Interferon in acute leukaemia in children. Lancet i: 166-167.

2. Cantell, K., L. Pyhala, and H. Strander. 1974. Circulating
   human interferon after intramuscular injection into animals
   and man. J. Gen. Virol. 22:453-455.

3. Desmyter, J., J. DeGroote, V.J. Desmet, A. Billiau, M.B. Ray,
   A.F. Bradburne, V.G. Edy, P. DeSomer, and J. Mortelmans.
   1976. Administration of human fibroblast interferon in
   chronic hepatitis B infection. Lancet ii: 645-647.

4. Greenberg, H.D., R.B. Pollard, L.I. Lutwick, P.B. Gregory,
   W.S. Robinson, and T.C. Merigan. 1976. Effect of human
   leukocyte interferon on hepatitis B virus infection in
   patients with chronic active hepatitis. New Engl. J. Med.
   295:517-522.

5. Jordan, G.W., R.P. Fried, and T.C. Merigan. 1974. Administration of human leukocyte interferon in herpes zoster: safety,
   circulating antiviral activity, and host responses to infection.
   J. Infect. Dis. 130:56-62.

6. Merigan, T.C., S.E. Reed, T.S. Hall, and D.A.J. Tyrrell.
   1973. Inhibition of respiratory virus infection by locally
   applied interferon. Lancet i: 563-567.

7. Strander, H., K. Cantell, G. Carlström, and P.A. Jakobsson.
   1973. Clinical and laboratory investigations on man:
   systemic administration of potent interferon to man. J.
   Natl. Cancer Inst. 51:733-742.

8. Cantell, K., S. Hirvonen, K.E. Mogensen, and L. Pyhala. 1974.
   Human leukocyte interferon: purification, stability and
   animal experiments. Pgs. 35-38 in C. Waymouth, ed. The
   Production and Use of Interferon for the Treatment and
   Prevention of Human Virus Infections. The Tissue Culture
   Association, Rockville.

9. Strander, H., K.E. Mogensen, and K. Cantell. 1975. Production
   of human lymphoblastoid interferon. J. Clin. Microbiol.
   1:116-117.

10. Havell, E.A., and J. Vilcek. 1972. Production of high-
    titered interferon in cultures of human diploid cells.
    Antimicrob. Ag. Chemother. 2:476-484.

11. Billiau, A., M. Joniau, and P. DeSomer. 1973. Mass produc-
     tion of human interferon in diploid cells stimulated by
     poly-I:C. J. Gen. Virol. 19:1-8.

12. Ho, M., J.M. Tan, and J.A. Armstrong. 1972. Accentuation
     of production of human interferon by metabolic inhibitors.
     Proc. Soc. Exp. Biol. Med. 139:259-262.

13. Bonin, O., I. Schmidt, K. Schmidt, R. Mauler, and H.
     Gruschkau. 1974. Calf serum content of cell culture grown
     viral vaccines and possibilities for its reduction. J.
     Biol. Standard. 2:139-141.

14. Stewart II, W.E., L.B. Gosser, and R.Z. Lockart. 1971.
     Priming : a non-antiviral function of interferon. J.
     Virol. 7:792-801.

15. House, W., M. Shearer, and N.G. Maroudas. 1972. Method for
     bulk culture of animal cells or plastic film. Exp. Cell.
     Res. 71:293-296.

16. Edy, V.G., I.A. Braude, E. DeClercq, A. Billiau, and P.
     DeSomer. 1976. Purification of interferon by adsorption
     chromatography on controlled pore glass. J. Gen. Virol.
     33:517-521.

17. Edy, V.G., M. Joniau, A. Billiau, and P. DeSomer. 1974.
     Stabilisation of mouse and human interferons by acid pH
     against inactivation due to shaking and guanidine hydrochlo-
     ride. Proc. Soc. Exp. Biol. Med. 146:249-253.

18. Havell, E.A., B. Berman, C.A. Ogburn, K. Berg, K. Paucker,
     and J. Vilcek. 1975. Two antigenically distinct species
     of human interferon. Proc. Nat. Acad. Sci. 72:2185-2187.

19. Stewart II, W.E., P. DeSomer, V.G. Edy, K. Paucker, K.
     Berg, and C.A. Ogburn. 1975. Distinct molecular species
     of human interferons : requirements for stabilization and
     reactivation of human leukocyte and fibroblast interferons.
     J. Gen. Virol. 26:327-331.

20. Edy, V.G., A. Billiau, and P.DeSomer. 1976. Human
     fibroblast and leukocyte interferons show different dose-
     response curves in assay of cell protection. J. Gen. Virol.
     31:251-255.

FACTORS INFLUENCING PRODUCTION OF INTERFERON BY HUMAN

LYMPHOBLASTOID CELLS

M. D. Johnston, Ph.D., K. H. Fantes, Ph.D., N. B. Finter,
M. B., B. Chir

Department of Virology Research & Development
Wellcome Research Laboratories
Langley Court, Beckenham, Kent, BR3 3BS

INTRODUCTION

Up till now, leukocytes obtained from blood donations have
been the main source of human interferon for clinical use.
Production from these has been undertaken almost solely by Cantell
and his colleagues in Finland, who have devised a highly
efficient process (1). However, to make a large amount of this
type of interferon entails that buffy coats are handled from many
individual donors, which is laborious and expensive, and makes
control relatively difficult. If interferon proves to be of
value in even some of the clinical trials that are at present
underway, then other sources will have to be found if the demand
for interferon is to be met.

One alternative is to make interferon from diploid fibroblast
cells, which under appropriate conditions can yield large amounts
of interferon (2, 3). High yielding cells can fairly readily be
obtained, and procedures for the control of production are
relatively straight forward. However, fibroblast interferon is
less stable than leukocyte interferon, and this could make it
more difficult to produce on a large scale. Also, leukocyte and
fibroblast interferons differ in other respects, for example in
their antigenicity (4) and in their inhibitory effects on the
growth of at least some cells (5). It is thus quite possible
that both types of interferon may be needed for different clinical
purposes.

We have been investigating another source of interferon,
namely lymphoblastoid cells, and in particular cells of the Namalva

61

line (6, 7). In all respects that we have investigated, interferon
derived from these cells has behaved identically to leukocyte interferon. It has the same iso-electric points and molecular weight, an
it appears to have the same stability to heat, pH and various chemic
treatments. In terms of biological properties, both are equally effective in inhibiting the growth of a number of viruses, and when us
at equivalent doses in terms of units of antiviral activity, have th
same inhibitory effect on the growth of various human cell lines.
Thus lymphoblastoid cells seem to represent an alternative source of
the leukocyte type of human interferon. This paper deals with some
aspects of the production of interferon from Namalva cells.

MATERIALS AND METHODS

Cells and Viruses

Namalva cells were obtained from Dr. Gresser, Villejuif. They
were grown in RPMI 1640 medium supplemented initially with 10% v/v
foetal calf serum, but now with 7% v/v serum from 6 - 15 month old
calves. The cells were grown in stationary cultures, e.g. 10 ml in
cm^2 tissue culture flasks or in 2 - 10 L volumes in stirred flasks
gassed with 5% CO_2 in air. The cells were subcultured every 2 to 3
days when the cell counts reached 2 - 3 x 10^6 cells/ml.

Cells of a continuous line of African Green monkey kidney, V_3(8
were grown in Eagle's Basal medium containing 5% v/v foetal calf
serum.

Sendai virus was propagated in fertile hen's eggs. Allantoic
fluid harvested three days after infection was clarified by light ce
trifugation and stored frozen in small samples at -70°C. Infectivit
and haemagglutinin (HA) assays were carried out by conventional
techniques.

Semliki forest virus was grown and assayed in chick embryo fibroblasts. For storage, 10% v/v foetal calf serum was added, and
the virus was kept at -70°C.

Induction of Interferon

The standard method used to induce interferon was as follows.
Cells were removed from growth medium by centrifugation at 750 x g.
The cell pellet was resuspended in RPMI 1640 medium containing 2% v/
calf serum, and the count adjusted to a concentration of 5 x 10^6
cells/ml. The cell suspension was distributed in 10 ml volumes into
plastic tissue culture flasks with a growth area of 25 cm^2, and the
cells were primed by adding lymphoblastoid interferon to a final
concentration of 50 - 100 units/ml.

The flasks were placed at 35°C for 2 h, and the cells were then induced to form interferon by adding Sendai virus at a final 20 - 50 HA units/ml. After a further 16 h at 35°C, the interferon was harvested by sedimenting the cells at 1500 x g. Acid was added to adjust the pH to 2.0 in order to inactivate the residual Sendai virus. After a minimum of 24 h, the pH was returned to 7.0, and the sample was assayed for interferon.

Interferon Assay

Assays were carried out by a modification of the dye-uptake method (9). Monolayer cultures of V_3 cells were grown on the bottom of glass vials of 20 mm internal diameter in foil covered racks in a humidified CO_2 incubator. 1 ml amounts of interferon dilutions were prepared in Eagle's Basal medium containing 2% v/v calf serum, and were placed on duplicate cell monolayers. The cultures were incubated overnight, and the cells were then challenged with approximately 50 plaque-forming units of Semliki forest virus. After incubation for a further 48 h, the extent of the cytopathic effect was determined indirectly by adding to each vial neutral red at pH 6.5 to a final concentration of 0.01%. After incubation for 1 h at 37°C, the excess stain was removed and the cell sheets were washed. The incorporated stain was eluted into 4 ml of a 1 : 1 mixture of ethanol and 0.1 M $NaH_2 PO_4$, and the concentration in each vial was measured at 545 nm in a color-imeter, by means of a fibre optics probe (Brinkmann, Westbury,N.Y.). A computer programme was employed to calculate the endpoints. A carefully calibrated internal standard interferon preparation was included in all assays, and results were calculated in terms of the human interferon research reference standard, 69/19 (with the method used, a vial of this standard was found to contain 1200 50% dye-uptake units, compared with its defined content of 5000 research reference units).

In routine assays in which interferon containing samples were diluted in 0.5 log_{10} steps, differences in interferon titres of 0.3 log_{10} or greater were statistically significant.

RESULTS

Screening of Lymphoblastoid Cell Lines

Dr. Christofinis in our laboratories has screened a large number of different lymphoblastoid cell lines for their ability to produce interferon when induced with Sendai virus. Half of the cells produced very little or no detectable interferon, but about 10% of the cells gave titres of 3000 units/ml. However, none of these proved superior to Namalva cells in terms of interferon yield and growth rate, and we have therefore worked almost entirely with Namalva cells. The experiments to be described in this paper have been undertaken to determine the best conditions for the production of interferon from these cells.

Factors Influencing Interferon Production

Inducers. A number of different viruses and other inducers of interferon have been tested with Namalva cells. Results obtained with some of these are shown in Table 1. The best inducer found has been Sendai virus. Two lentogenic Newcastle disease virus vaccine strains, La Sota and B_1 gave rise to less or very much less interferon than Sendai virus, and low titres were also obtained in other experiments with six other strains of Newcastle disease virus (strains Texas, Herts, Lamb, AG68V, Queensland and Ulster). Semliki forest virus, which replicates in Namalva cells (J. Morser, personal communication), gave rise to no detectable interferon. Poly rI. poly rC when combined with DEAE-dextran is a potent inducer of interferon in human fibroblasts. However, this combination gave rise to only small amounts of interferon in Namalva cells. It has recently been suggested that interferon may be induced in consequence of inhibition of host cell protein synthesis (10), but one such inhibitor, cycloheximide, did not stimulate interferon production.

Table I

Interferon Yields from Namalva Cells using Different Inducers

Inducer	Amount added	Interferon Titre (Log_{10} units/ml)
Sendai virus	100 HAU/ml	4.85
NDV La Sota	50 HAU/ml	4.13
NDV Bl	50 HAU/ml	3.32
Semliki Forest virus	5 pfu/cell	<2.00
Poly rl : rC + DEAE - Dextran*	100 µgr/ml	2.02
Cycloheximide†	30 µgr/ml	<2.00

* A mixture of 100 µgr/ml Poly rl : rC and 100 µ.gr/ml DEAE - Dextran

† Present from 0 to 8 hours

Multiplicity of Infection. We studied the amounts of Sendai
virus required to give maximum yields of interferon. The data
in Fig. 1 show that as little as 5 - 10 HA units/ml were sufficient
to give rise to maximum formation of interferon. This corresponds
to an added multiplicity of infection of 5 - 10 egg ID50/cell.

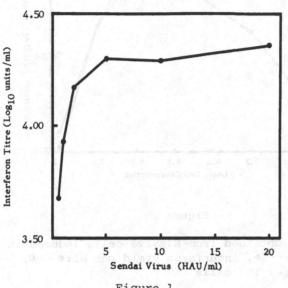

Figure 1

Effect of the multiplicity of infection with Sendai virus on
interferon yields.

Cell Concentration. Yields of interferon from cultures induced
at different cell concentrations are shown in Fig. 2, expressed
both in terms of interferon titre and interferon yield per
million cells. It will be seen that the optima are not identical
for the two methods of expressing interferon yields. From the
practical point of view, it is desirable to obtain the interferon
at the greatest possible concentration, since this makes subsequent
concentration and purification steps simpler. Therefore a cell
concentration of 5.0×10^6 cells per ml was selected for routine
use: at this concentration, the yield of interferon per million
cells is high, and titres per millitre are maximum.

At cell concentrations greater than about 8 million per ml,
there was a considerable decrease in interferon titres. Also, at
cell concentrations below about 3 million per ml, there was a
decrease in interferon yields in terms of yield per cell, and
particularly in terms of titre.

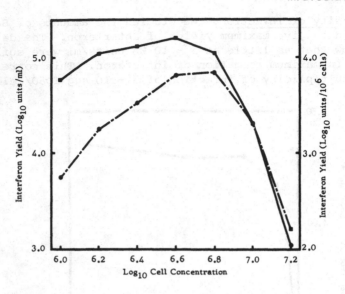

Figure 2

Interferon yields obtained from Namalva cells induced at different
concentrations. ●- - - ●, interferon yield per ml:●—●,
interferon yield per 10^6 cells

Temperature. Results of inducing interferon at different
temperatures are shown in Table 2. The titres obtained were maximum
at 35°C, and substantially lower if the cells were incubated at
33° or 39°C.

Table 2

Optimum Temperature for Interferon Production
by Namalva Cells

Temperature (°C)	Interferon Titre (Log_{10} units/ml)
33	3.75
35	4.53
37	4.26
39	3.52

Serum. The importance of serum during the induction process
was investigated. Cells were washed three times in phosphate
buffered saline to remove any residual serum carried over from the
growth medium. The cells were resuspended in serum-free RPMI
1640 medium, and foetal calf serum was added to the desired
concentrations. In six such experiments, no consistent difference
in interferon titres was found with cultures containing concentrations of foetal calf serum ranging from 0% to 5%. However, there
are some indications that the interferon produced in the absence
of serum is less stable. It should be noted that even when washed
in the way described, some serum is shed from cells into serum-
free medium during the induction period.

In several experiments, calf serum was omitted from the
induction medium, and replaced with human albumin fraction (saline)
B.P. Under these conditions the titres and stability of the interferon produced were the same as in control cultures with calf
serum.

Priming. Addition of interferon to Namalva cells 2 h prior
to induction with Sendai virus led to a small increase in interferon
yields. Results of a representative experiment are shown in Table 3.
The differences are small, but have been found in a number of similar
experiments. Maximum increases in yields are obtained when about
10 units/ml of interferon are used for priming. The duration of
the priming period is not important: the same yields were obtained
when cells were primed for different times up to 5 h before induction
with Sendai virus, but with a priming period of 19 h, much less
interferon was obtained.

Kinetics. A number of replicate cultures of Namalva cells
were induced simultaneously with Sendai virus. At different
times after induction, the interferon was harvested and assayed.
Results are shown in Fig. 3. Interferon production could be
detected by 4 h after induction, and yields increased steeply to
reach a maximum at about 10 h.

Superinduction. When fibroblasts are induced to form
interferon by the addition of Poly rI : poly rC, yields can be
substantially augmented by the appropriate addition of the metabolic
inhibitors, cycloheximide and actinomycin D (11, 12). We therefore
examined the effects of these substances on the production of
interferon by Namalva cells induced with Sendai virus. Actinomycin
D was added for 1 h at various times after induction, and other
cultures were treated with cycloheximide from 3 till 10 h and with
actinomycin D from 9 till 10 h after induction. In all instances,
the interferon was harvested 36 h after induction. Table 4 shows
results obtained. Treatment of cells with actinomycin D at 4,
6 or 8 h after induction stopped further interferon formation
(compare Fig. 3). Thus far from increasing interferon yields as

Table 3

Effect of Priming Namalva Cells on Interferon Yields

Interferon Added (units/ml)	Interferon Titre (Log_{10} units/ml)
0	4.14
3.2	4.26
10	4.40
32	4.41
100	4.29
320	4.42
1000	4.34

Priming Interferon added at time 0 hr.
Inducer added at time 2 hr.

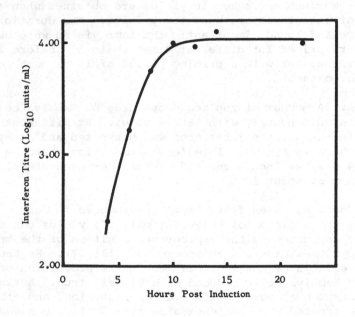

Figure 3

Kinetics of interferon production by Namalva cells induced with Sendai virus.

Table 4

Superinduction of Sendai Virus Induced Namalva Cells

Cycloheximide concentration (μgr/ml)	Time of Cycloheximide treatment (hr)	Actinomycin D concentration (μgr/ml)	Time of Actinomycin D treatment (hr)	Interferon titre (Log$_{10}$ units/ml)
0	-	0	-	4.31
0	-	2	4- 5	3.49
0	-	2	6- 7	4.07
0	-	2	8- 9	4.18
0	-	2	10-11	4.23
0	-	2	12-13	4.26
30	3 - 10	5	9-10	2.69

was found in the fibroblast system, actinomycin D apparently
blocks formation of interferon by Namalva cells. Also, the
use of both cycloheximide and actinomycin D greatly reduced
the final interferon titre. We have also attempted to increase
the titres of interferon by using UV-irradiated or heat-
inactivated Sendai Virus, but decrease titres were obtained.

  <u>Consistency of Production</u>. Samples from a subline of
Namalva cells were induced to form interferon under the
standard conditions described in the Methods section at
intervals during an 80 day period. The same calf serum,
inducing virus and medium (freshly prepared from concentrates)
were used throughout. As shown in Fig. 4 there was a difference
of more than ten-fold between the highest and the lowest interferon
titres obtained over this period. The reasons for these differences
are not known but similar results have been found in studies with
other sublines and in other laboratories (J. Morser, personal
communication). Thus, some other and as yet unknown factor or
factors play a part in determining the yield of interferon from
Namalva cells.

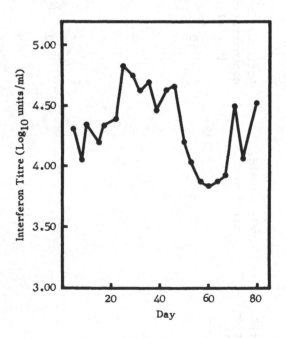

Figure 4

Interferon titres obtained on successive occasions from a single
subline of Namalva cells induced under standard conditions.

DISCUSSION

We have investigated a number of factors which influence the formation of interferon by Namalva cells, and have found those optimum for production in 25 cm^2 culture flasks. These do not necessarily apply when the cells are grown in other systems, for example in stirred flasks, or in large tanks. Thus conditions must by systematically investigated at each level of production. On the whole, titres from stirred or agitated cultures have tended to be slightly lower than those obtained from stationary cultures, for reasons which are not yet clear. Nevertheless, when grown on a large scale such as a 100 litre tank, Namalva cells are a good source of interferon and potentially they could provide large amounts of comparatively inexpensive human interferon of the leukocyte type.

It is interesting to compare the conditions which we have found optimal for producing interferon from Namalva cells with those described by Cantell and his associates for the production of interferon from primary leukocytes. With the latter cells, Sendai virus and the Victoria strain of Newcastle disease virus were found the best inducers (13, 14). We have found Sendai virus also to be very effective with Namalva cells, but most strains of Newcastle disease virus, including a strain, Herts, which is virulent like the Victoria strain, gave rise to much less interferon. The minimum amount of Sendai virus required for full yields of interferon were 5 - 10 HA units/ml, which is considerably less than the value of 100 - 300 HA units/ml recommended for leukocytes (13). However, leukocytes are induced at a concentration of 10×10^6 cells per ml, whereas Namalva cells were used at 5×10^6 cells per ml. It is also possible that the haemagglutination assays used in Helsinki and in our laboratories differ in their sensitivities. Thus possibly a less ambiguous way of expressing the required multiplicity of infection is in terms of egg ID50 doses per cell, and here our minimum is about 5 - 10 egg ID50 per cell. In practice we have usually used rather more than this in order to err on the safe side, and similarly we have used more than the bare minimum amount of priming interferon. In terms of the kinetics of interferon production, the optimum induction temperature and the yields of interferon, the results obtained with Namalva cells in our system are very similar to those published for the leukocyte system (15). We have also found that Namalva interferon can be purified by procedures very similar to those established for leukocyte interferon, including precipitation and extraction with organic solvents. This is hardly surprising in view of the close similarity or possibly identity of the two types of interferon. Indeed from this point of view, the concentration and quality of the contaminating proteins present are probably more important than the interferons themselves.

Perhaps the biggest difference between the production of
interferon with leukocytes and lymphoblastoid cells is the finding
that very little if any added serum protein need be present in the
medium during the period of interferon induction from Namalva
cells. In contrast, primary leukocytes make considerably less
interferon in the absence of added serum (16). This feature of
the Namalva cell induction system could have considerable practical
advantages: it may be possible to grown cells on a very large
scale in a medium containing an animal serum, e.g. calf serum,
and to wash the cells thoroughly to remove nearly all of this
serum before they are induced to form interferon. Provided that
the residual amounts of calf serum can be removed during the
purification process, this may eliminate the need to use human
serum when producing this type of interferon.

REFERENCES

1. Cantell K, Hirvonen S, Mogensen KE, et al. Human leukocyte
    interferon: Production, purification, stability, and animal
    experiments. The production and use of interferon for the
    treatment and prevention of human virus infections. Edited by
    C. Waymouth. Rockville, Tissue Culture Association, 1974,
    pp 35 - 38.
2. Havell EA, Vilcek J. Production of high titered interferon
    in cultures of human diploid cells. Antimicrob Agents
    Chemother 2: 476 - 484, 1972.
3. Billiau A, Joniau M, DeSomer P. Mass production of human
    interferon in diploid cells stimulated by poly I : C. J. Gen
    Virol 19: 1 - 8, 1973.
4. Havell EA, Berman B, Ogburn CA, et al. Two antigenically
    distinct species of human interferon. Proc Nat Acad Sci USA
    72: 2185 - 2187, 1975
5. Einhorn S, Strander H. Is interferon tissue specific? -
    Effect of human leukocyte and fibroblast interferons on the
    growth of lymphoblastoid and osteosarcoma cell lines.
    J. Gen Virol 1977, in press.
6. Klein G, Dombos L, Gothoskar B. Sensitivity of Epstein-
    Barr virus (EBV) producer and non-producer human lymphoblastoid
    cell lines to superinfection with EB virus. Int J Cancer
    10: 44 - 57, 1972.
7. Strander H, Mogensen KE, Cantell K. Production of human
    lymphoblastoid interferon. J Clin Microbiol 1: 116 - 117,
    1975.
8. Christofinis GJ.: Biological characteristics of the cell line
    GL - V3 derived from the kidney of a vervet moneky
    (Cercopithecus Aethiops). J Med Microbiol 3 : 251 - 258, 1970.
9. Finter NB. Dye-uptake methods of assessing viral cytopatho-
    genicity and their application to interferon assays. J Gen
    Virol 5: 419 - 427, 1969.
10. Tan YH, Berthold W. : A mechanism for the induction and
    regulation of human fibroblastoid interferon genetic expression.
    J Gen Virol 34: 401 - 411, 1977.
11. Tan YH, Armstrong JA, Ke YJ, et al. Regulation of cellular
    interferon production : Enhancement by antimetabolites.
    Proc Nat Acad Sci USA 67: 464 - 471, 1970.
12. Vilcek J, Ng MH Post-transcriptional control of interferon
    synthesis. J Virol 7: 588 - 594, 1971.
13. Strander H, Cantell K. Production of interferon by human
    leukocytes in vitro. Ann Med Exp Biol Fenn 44: 265 - 273,
    1966.
14. Strander H, Cantell K. Further studies on the production of
    interferon by human leukocytes in vitro. Ann Med Exp Biol
    Fenn 45 : 20 - 29, 1967.

15. Tovell D, Cantell K.   Kinetics of interferon production in
     human leukocyte suspensions. J Gen Virol 13:   485 - 489, 1971.
16. Hadbrazy G, Strander H, Cantell K.   Serum requirement for
     interferon production by suspended human leukocytes:
     Studies on action of serum. J Gen Virol 5:: 351 - 358, 1969.

ANTIGENIC PROPERTIES AND HETEROSPECIFIC ANTIVIRAL ACTIVITIES OF

HUMAN LEUKOCYTE INTERFERON SPECIES

K. Paucker, Ph. D., B.J. Dalton, B.S., and E.T. Torma, M.S.

Microbiology Department, The Medical College of Penn-

sylvania, 19129

ABSTRACT

Human interferon obtained in peripheral leukocytes was purified approximately 1000-fold by affinity chromatography on anti-leukocyte interferon globulins coupled to Sepharose 4B, and by filtration on SDS-Sephadex G-100. The interferon was subsequently resolved into two molecular species by adsorption chromatography on SDS-hydroxylapatite. The two species which were eluted at different phosphate molarities from hydroxylapatite, could also be distinguished on the basis of electric charge properties and they migrated at different rates in SDS-polyacrylamide gels. Crossneutralization tests with monospecific rabbit anti-leukocyte and anti-fibroblast interferon sera revealed that the two species possessed leukocyte interferon-specific antigenic determinants. Both were immunogenic in mice and they were neutralized to a comparable degree by antisera against either component. A variable degree of antiviral activity was expressed by both interferon components in bovine, porcine and murine cells. However, the two interferon species were equally active in this respect, and the protective effects exhibited in homologous and heterologous cell cultures were similarly susceptible to reduction by beta-mercapto-ethanol. We conclude that the two molecular species of human leukocyte interferon are biologically similar.

INTRODUCTION

Examination of human leukocyte interferon preparations by a variety of techniques has revealed the existence of two dominant

75

molecular populations (1-6). Aside from the reported differences
in apparent molecular weights (7) and, when electrophoresed in the
presence of a reducer, the antiviral activity of one of the components in human cells was completely abolished, whereas protective
activity in cat cells was fully retained. Under the same conditions,
the other component remained active in both types of cell cultures
(8). These observations suggested that the two molecular species
in human leukocyte interferon possessed remarkable distinctive
physiocochemical and biological features which might clearly
influence approaches toward its purification. Therefore, we de-
cided to seek more information on the biological properties associated with these components. Efforts were primarily with these
components. Efforts were primarily focused on two arease, i.e.,
antigenicity and protective effects in cells from other hosts.
Furthermore, an attempt was made to study the isolated interferon
components in somewhat purified form so that the possible participation of extraneous factors in the events studied could be some
extent be minimized. A brief description of our observations and
conclusions follows below, since the major findings have been published elsewhere (9).

The human leukocyte interferon used in these studies was prepared
at the Finnish Red Cross Blood Transfusion Service, Helsinki,
Finland, as previously reported (10,11). The concentrated lyo-
philized product contained 1×10^3 reference units per mg protein.
The dissolved interferon was then purified by a series of steps
which included affinity chromatography on anti-leukocyte interferon
globulins coupled to Sepharose 4B (12), followed by treatment
with SDS and gel filtration through Sephadex G-100 in the presence
of SDS (5). Eventually, the interferon was chromatographed on SDS-
hydroxylapatite and resolved into two components by elution in a
gradient of rising phosphate molarity and at constant pH. Component A, which eluted at the lower phosphate molarity estimated at
0.21, corresponded to the 21,000 molecular weight species of Stewart
and Desmyter (7), and component B, eluting at a calculated phosphate molarity of 0.33, to their more rapidly migrating interferon
species. A detailed description of these procedures and relation-
ships has been provided in an earlier report (5).

Human leukocyte interferon was previously shown to contain fibroblast interferon-specific activity (12, 13), and, in fact, such
a factor could be separated and isolated from leukocyte interferon
by affinity chromatography on anti-fibroblast interferon globulins
bound to a Sepharose carrier (14, 15). Therefore, we examined the
two interferon species in crossneutralization tests with monospecific
rabbit antisera which neutralized either leukocyte or fibroblast
interferon, but not both, to see whether one or the other component may be fibroblast-specific. The results shown in Figure 1 and
modified from experiments published earlier (9) demonstrate that
leukocyte interferon components A and B, as well as a mixed

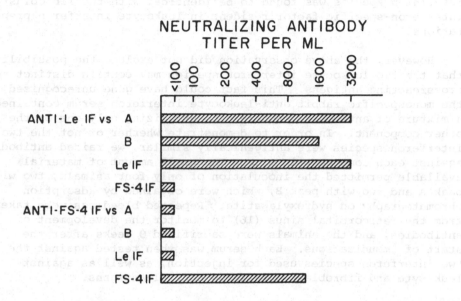

Figure 1

Leukocyte interferon specificity of A and B components in human
leukocyte interferon. Antigens are peaks A and B, separated on
SDS-hydroxylapatite and SDS-polyacrylamide gel electrophoresis,
as well as leukocyte (le) and fibroblast (FS-4) interferon (IF)
controls. Neutralization was carried out in microplates against
8 reference units of each antigen, and vesicular stomatitis virus
was used for challenge. Presentation modified from Paucker et al.
(9).

leukocyte interferon control, were neutralized only by anti-leukocyte interferon serum. The converse is true when the same antigens

are tested against the anti-fibroblast interferon serum. In that
case, only the fibroblast interferon control but none of the
other antigens were neutralized. Therefore, neither of the two
interferon species was found to be identical with the fibroblast
interferon-specific factor resident in leukocyte interferon preparations.

However, the above observation did not exclude the possibility
that the two leukocyte interferon species may contain distinct non-
crossreacting antigens. This fact could have gone unrecognized if
the monospecific rabbit anti-leukocyte interferon serum contained
a mixture of antibodies capable of neutralizing only one or the
other component. In order to demonstrate whether or not the two
interferon species were antigenically similar, we raised antibodies
against each component in mice. The small amount of material
available permitted the inoculation of only four animals, two with
peak A and two with peak B, which were obtained by adsorption
chromatography on hydroxylapatite. Repeated bleedings were taken
from the retroorbital sinus (16) to monitor the development of
antibodies, and the animals were sacrificed 9 weeks after the
start of immunizations. Each serum was then tested against the
two interferon species used for injection, as well as against
leukocyte and fibroblast interferon control antigens.

The results obtained with sera from one animal in each group,
collected after 8 and 9 weeks, respectively, are illustrated in
Figure 2. There were no significant differences observed between
the two antisera. Each serum neutralized components A and B, as
well as the leukocyte interferon control antigen, equally well.
However, unlike in hyperimmunized animals, (12, 17), neither serum
contained antibodies against fibroblast interferon. Presumably,
this may have been due to the small quantity of fibroblast-specific
interferon in the immunizing inoculum. Alternatively, the fibroblast-specific interferon factor may have been eliminated by the
SDS-treatment, or any of the ensuing purification and separation
steps. However, the main conclusion to be drawn from the experiment is that there are no major antigenic differences detectable
between the two leukocyte interferon molecular populations.

The reported differences in antiviral activities of the two
leukocyte interferon species in rabbit (7) and cat cells (8)
prompted us to look at other types of heterologous cultures in
which leukocyte interferon was previously shown to be active,
namely mouse (4, 18), as well as bovine and porcine cells (19).
Initially we had also included rabbit RK-13 cells, but in our hands
SDS-treatment abolished the protection by the isolated, purified
components (as well as of rabbit interferon) in these cells. Titrations were conducted with interferon components A and B obtained
from hydroxylapatite, and with crude leukocyte interferon, in
human, mouse, swine and bovine cells. The results shown in

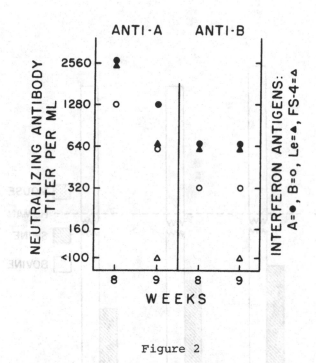

Figure 2

Antigenic similarity of A and B components in human leukocyte
interferon. Antisera were obtained in random bred Swiss mice bled
after 8 and 9 weeks of weekly subcutaneous injections of 60,000
interferon units each of peaks A or B, isolated on SDS-hydroxylapatite. Neutralization tests performed as in Figure 1. Modified
from Paucker et al. (9).

Figure 3 indicate that all three types of interferon preparations
behaved similarly in the cell systems studied. Moreover, the
heterospecific protective effects observed corresponded quite
well with those noted previously. In mouse cells, the detected
antiviral activity was less than 1% of that found in human cells
(18), but it was clearly measurable and amounted to between 300
and 500 units, as compared to 100,000 units in human fibroblasts.
The protective activity in porcine cells was intermediate though
also considerably less than in human cells. On the other hand, the
titer in bovine cultures was from 20 to 30-fold higher than in
human fibroblasts (19).

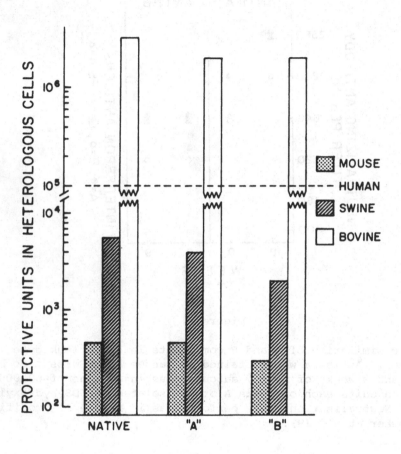

Figure 3

Antiviral activities of human leukocyte interferon components in
mouse, swine and bovine cells. Titers were corrected on the basis
100,000 units in human cells as indicated by the broken line.
Details regarding origin of the cells and assay systems were given
elsewhere (9). Components A and B were obtained separately by
hydroxylapatite chromatography; the native preparation was non-
purified human leukocyte interferon.

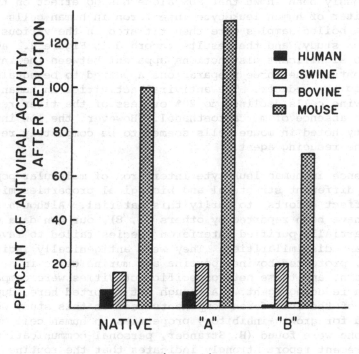

Figure 4

Effect of beta-mercaptoethanol on protective activities of human
leukocyte interferon components in human and other mammalian cells.
Treatment was for 1 min at 100°C in the presence of 1% SDS, 5 M-
urea and 1% beta-mercaptoethanol. Titers of controls treated with
SDS alone were taken as 100%. Interferon components were obtained
by SDS-polyacrylamide gel electrophoresis of peaks A and B isolated
from hydroxylapatite. Native interferon consisted of the non-
purified product from peripheral leukocytes. Further details
were provided elsewhere (9).

Since these data also indicated that there were no major dis-
cernible differences between the two leukocyte interferon species
as far as their activities in heterologous cells were concerned, we
examined next the impact of mercaptoethanol on these effects.
Components A and B as well as a crude leukocyte interferon, containing a mixture of the two species, were each divided into two portions, one being treated with SDS alone, the other receiving in addition mercaptoethanol and urea, as specified in the legend of Figure 4.

It had previously been shown that SDS alone has no effect on the
interferon titer of human leukocyte interferon in human cells
(7, 20). The boiled samples were then titrated in the various cell
cultures under study, and the results recorded in Figure 4. Again,
there were no significant distinctions apparent between the interferon components. The three preparations appeared to be equally
susceptible to the reducer. The antiviral activities in human,
swine and bovine cells declined to 20% or less of the titer recorded in the absence of mercaptoethanol. However, the low interferon activity noted in mouse cells seemed to be completely resistant to the reducing agent.

The presence in human leukocyte interferon of molecular populations with different structural and biological properties might
profoundly affect efforts to purify this material. Although such
differences have been reported by others (7, 8), our own data with
individual partially purified interferon species failed to reveal
any significant dissimilarities. They were antigenically indis-
tinguishable, protected bovine, porcine and murine cells in a
similar pattern, and these heterospecific activities were comparably
affected by a reducing agent. Although not reported here, duplicate smaples of the interferon components used in this study were
also examined for growth-inhibitory properties in human cell cultures, and none were found (H. Strander, personal communication).
Moreover, a recent report strongly indicates that the routine
practice of acidifying interferon preparations to inactivate
inducer virus may profoundly affect the distribution of leukocyte
interferon components (E. Törmä, personal communication). We,
therefore, conclude that the two leukocyte interferon molecular
species are essentially similar and need not interfere with
attempts to purify this interferon.

ACKNOWLEDGMENT

Supported in part by contract NO1-AI-52520 from the National
Institute of Allergy and Infectious Diseases. The generous
contribution of the Finnish Red Cross Blood Transfusion Service
to this program is gratefully acknowledged.

REFERENCES

1. Fantes, K.H. 1969. Partial purification, concentration and properties of human leukocyte interferon. L'Interferon, Colloques de L'Institut National de la Santé et de la Recherche Medicale No. 6. Pgs. 181-186.

2. Matsuo, A., S. Hayashi, and T. Kishida. 1974. Production and purification of human leukocyte interferon. Japan J. Microbiol. 18:21-27.

3. Anfinsen, C.B., S. Bose, and L. Corley et al. 1974. Partial purification of human interferon by affinity chromatography. Proc. Nat. Acad. Sci. USA 71:3139-3142.

4. Borecký, L., N. Fuchsberger, and V. Hajnicka. 1974. Electrophoretic profiles and activities of human interferon in heterologous cells. Intervirology 3:369-377.

5. Törmä, E.T. and K. Paucker. 1976. Purification and Characterization of human leukocyte interferon components. J. Biol. Chem. 251:4810-4816.

6. Chen, J.K., W.J. Jankowski and J.A. O'Malley et al. 1976. Nature of the Molecular Heterogeneity of Human Leukocyte Interferon. J. Virol. 19:425-434.

7. Stewart, W.E., II, and J. Desmyter. 1975. Molecular heterogeneity of human leukocyte interferon: two populations differing in molecular weights, requirements for renaturation, and cross-species antiviral activity. Virology 67:68-73.

8. Desmyter, J. and W.E. Steward II. 1976. Molecular modification of interferon: attainment of human interferon in a conformation active on cat cells but inactive on human cells. Virology 70:451-458.

9. Paucker, K., B.J. Dalton, and E.T. Törmä et al. 1977. Biological properties of human leukocyte interferon components. J. Gen. Virol. 35:341-351.

10. Strander, H. and K. Cantell. 1966. Production of interferon by human leukocytes _in vitro_. Annales Medicinae Experimentalis et Biologiae Fenniae 44:265-273.

11. Cantell, K. 1970. Preparation of human leukocyte interferon. Pgs. 6-8 in F.T. Perkins and R.H. Regamey, eds. International Symposium on Interferon and Interferon Inducers. Symposia Series in Immunobiological Standardization. Vol. 14. Karger, Basel-New York.

12.   Berg, K., C.A. Ogburn, and K. Paucker et al. 1975. Affinity
      chromatography of human leukocyte and diploid cell interferons
      on sepharose-bound antibodies. J. Immunol. 114:640-644.

13.   Havell, E.A., B. Berman, and C.A. Ogburn et al. 1975. Two
      antigenically distinct species of human interferon. Proc.
      Nat. Acad. Sci. USA 72:2185-2187.

14.   Paucker, K., B.J. Dalton, and C.A. Ogburn et al. 1975.
      Multiple active sites on human interferons. Proc. Nat. Acad.
      Sci. USA 72:4587-4591.

15.   Havell, E.A., B. Berman, and J. Vilček. 1975. Antigenic
      and biological differences between human leukocyte and
      fibroblast interferons. Pgs. 49-61 in Proceedings of the
      Symposium on the Clinical Use of Interferon, Yugoslav Academy
      of Sciences and Arts, Zagreb.

16.   Riley, V. 1960. Adaptation of orbital bleeding technic to
      rapid serial blood studies. Proc. Soc. Exp. Biol. Med. 104:
      751-754.

17.   Paucker, K., C.A. Ogburn, and K. Berg. 1975. Purification
      of interferons by immune affinity chromatography. Pgs. 76-84
      in T. Hasegawa, ed. Proceedings of the 1st Intersectional
      Congress of the International Association of Microbiological
      Societies. Vol. 4. Science Council of Japan.

18.   Levy-Koenig, R.E., R.R. Golgher and K. Paucker. 1970.
      Immunology of interferons. II. Heterospecific activities of
      human interferons and their neutralization by antibody. J.
      Immunol. 104:791-797.

19.   Gresser, I., M-T. Bandu, and D. Brouty-Boye et al. 1974.
      Pronounced antiviral activity of human interferon on bovine
      and porcine cells. Nature 251:543-545.

20.   Mogensen, K.E. and K. Cantell. 1974. Human leukocyte inter-
      feron: a role for disulphide bonds. J. Gen. Virol. 22:95-103.

MEMBRANE ALTERATIONS FOLLOWING INTERFERON TREATMENT

E.H. Chang, Ph.D., E.F. Grollman, M.D.*, F.T. Jay, Ph.D.,
G. Lee, M.D.*, L.D. Kohn, M.D.* and R.M. Friedman, M.D.

Laboratory of Experimental Pathology and
*Laboratory of Biochemical Pharmacology, National
Institute of Arthritis, Metabolism, and Digestive
Diseases, National Institutes of Health, Bethesda,
Maryland 20014

ABSTRACT

Interferon treatment appears to induce a number of changes in
the plasma membrane of uninfected cells. Interferon treatment
altered the surface exposure of gangliosides of both L_y and KB cell
membranes. The differences were found in the amount and pattern
of incorporation of tritium after galactose oxidase treatment. In
AKR,C⁻ (AKR-2B) mouse cells, not only was there an apparent increase
in the number of intramembranous particles in reponse to treatment
with interferon but also the kinetics of the increase followed that
of the establishment of the antiviral activity. The buoyant
density of plasma membrane was also found to be significantly increased in interferon-treated cells. Moloney murine leukemia virus
produced in interferon-treated mouse thymus and bone marrow cells
had a high particle to infectivity ration. This virus contained
a prominent glycoprotein with a molecular weight of about 85,000.
This large glycoprotein was only a very minor component of Moloney
leukemia virus produced in control TB cells and might be an
uncleaved precursor to gp 69-71.

INTRODUCTION

The notion of a cell surface receptor specific for interferon
was suggested in earlier studies (1, 2). Several reports indicated
that the oligosaccharide portion of membrane glycolipids (gangliosides) was involved in the interaction of interferon with the cell

membrane (3, 4, 5, 6). The finding that preincubation with phytohemagglutinin (PHA) blocked interferon action suggested that the
PHA-binding site on the cell membrane might be involved in interferon action (7). Preincubation of human interferon with gangliosides was shown to inactivate the antiviral effect of interferon
(8).

A similarity in mechanisms between interferon's initiation of
an antiviral state and the transmission of messages through the
cell membrane by thyrotropin (TSH) and cholera toxin was suggested
(6, 9, 10, 11). The interaction between these substances and
surface receptors was likely to be the initial step(s). Several
other changes in the cell surface brought about after interferon
binding have been reported. Mouse interferon has been shown to
enhance the expression of cell surface histocompatibility antigens
in L 1210 cells (12). The negative charge on the surface of L_y
cells was also increased by treatment with interferon resulting
in the cells having higher electrophoretic mobility toward the
anode (13). In a recent paper, it was demonstrated that interferon
treatment of AKR,C$^-$ cells led to establishment of an antiviral
state and concomitant alterations in the cell plasma membrane. An
increase in the number of intramembranous particles was observed in
freeze fracture electron micrographs and the density of plasma
membrane preparations was significantly altered (14). In L_y cells,
it was shown that interferon treatment altered the cell surface
exposure of gangliosides, changed the electrochemical gradient of
the cell membrane, and increased the levels of cyclic AMP in the
cell (6, 15).

The correlation between the cell surface alteration and establishment of an antiviral state has been proposed (1, 13, 14).
However, it was found that alteration in membrane after interferon
treatment was not sufficient in all cases to result in the transmission of a message to start the production of antiviral protein
in the cell (15). This could reflect the differences in sensitivity
found in various cell lines to interferon treatment. For example,
interferon-insensitive DB cells had slightly altered glycoproteins
after interferon treatment; however, no antiviral state was induced
in KB cells.

These data together with the results reported previously from
this and other laboratories have suggested the antiviral activity
of interferon may be expressed through induction of alterations
in the cell membrane. This may also explain the decrease in infectivity of murine leukemia viruses which are produced by interferon-treated cultures of TB and by some AKR cell cultures (5, 14,
16, 17).

In wild type Moloney leukemia virus infected TB cells, virus
particles defective in infectivity were produced (16, 17). The

low infectivity might be due to the increase in the fraction of
a large glycoprotein (85K) in the virions (18). This glycoprotein
(gp 85) may be a uncleaved precursor of gp 69/71 (19, 20).

RESULTS AND DISCUSSION

Interferon treatment appears to bring about a number of changes
in plasma membrane preparations of L_y cells exposed to interferon
developed a complex alteration in their ability to bind cholera
toxin. At low interferon concentrations binding of toxin was
significantly increased. When the interferon concentration was
increased, however, binding was inhibited (10).

Effect of Interferon on the Composition and Surface
Exposure of Gangliosides and Other Components in the
Membranes of Mouse L Cells

$\underline{N}$-acetylgalactosaminyl-[$\underline{N}$-acetyl-neuraminyl]-galactosylgluco-
sylceramide (G_{M2}) and $\underline{N}$-acetylneuraminylgalactosylglucosylceramide
(G_{M3}), predominant gangliosides in the plasma membranes of inter-
feron-sensitive L_y cells (Figure 1A), have been implicated as interferon-specific receptor components or receptor analogs (3, Dr. C. B.
Affinsen, personal communication). There was no difference in
ganglioside pattern which could be detected in membranes isolated
from L_y cells before or after interferon treatment (Figure 1B) or
when labeled with [^{3}H]-sodium borohydride after mild sodium metaperiodate treatment. However, there were differences found in the
amount and pattern of incorporation of tritium after galactose
oxidase treatment. The decrease in [^{3}H]-incorporation in interferon-treated membranes indicated that less of the oligosaccharide
moieties were exposed on the surface of the cell membrane (Figure
2A and B). This might be due to the change in orientation of these
gangliosides in the membrane. Since galactosyl-$\underline{N}$-acetylgalacto-
saminyl-[$\underline{N}$-acetylneuraminyl]-galactosylglucosylceramide (G_{M1}) and
galactosyl-$\underline{N}$-acetylgalactosaminyl-[$\underline{N}$-acetylneuraminyl-N-acetylneura-
minyl]-galactosylglucosylceramide ($\overline{G}_{D1b}$) have been implicated as
components of the receptors for cholera toxin and TSH respectively
(11), this result was consistent with the previous observations
that pretreatment of L_y cells with interferon altered the binding
of both TSH and cholera toxin.

Interferon treatment decreased cell surface exposure of the
terminal galactose residue in oligosaccharides of four membrane
glycoproteins. Changes were also found in cell surface exposure of
the neutral glycolipids of L_y cell membrane including a decrease
in some components and an increase in other components.

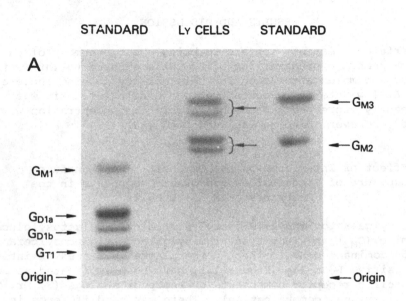

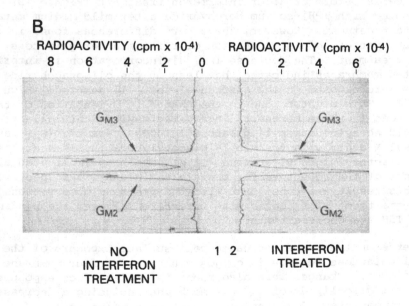

Figures 1A & 1B

Figure 1A.   Thin-layer chromatography of gangliosides extracted
from L_y cell membranes.   Gangliosides are visualized with
resorcinol spray.   The standards in both cases are gangliosides
isolated from bovine brain.

Figure 1B.   Radioscans of gangliosides extracted from the membranes
of L_y cells which had been treated with interferon (lane 2) and
gangliosides extracted from the membranes of L_y cells which had
never been exposed to interferon (lane 1).   Interferon treatment
was overnight with 100 unit/ml; the membranes were sequentially
treated with sodium metaperiodate and [3H]-labeled sodium borohydride.   120,000 and 104,000 cpm were applied to lanes 1 and 2,
respectively.   After radioscans were performed, the gangliosides
on each plate were visualized with resorcinol.

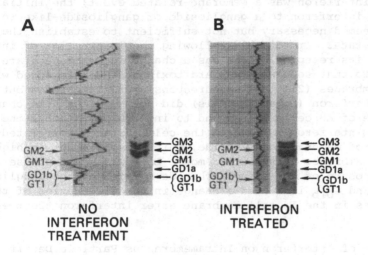

Figure 2A & 2B

Radioactive scans of gangliosides extracted from L_y cells whose
membranes had been sequentially treated with galactose oxidase and
[3H]-labeled sodium borohydride, i.e., of gangliosides whose terminal galactose residues had been tritiated.   (A) Gangliosides from L_y
cells which had never been treated with interferon.   (B) Gangliosides
from L_y cells which had been treated overnight with 100 units/ml of
interferon.   Equal amounts of the gangliosides from the interferon-
treated (B) and untreated (A) cells were applied to the thin-layer
scans as seen by the resorcinol-stained pattern next to each scan.
The radioactivity applied to the plate in A was 10,000 cpm, in
B 1,000 cpm.

Effect of Interferon on the Composition and Surface
Exposure of Gangliosides and Other Components in the
Plasma Membrane of Human KB Cells

KB cells nonresponsive to interferon but able to bind interferon differed from the L_y systems. The total ganglioside content per mg plasma protein was one-eighth of that found in L_y cell. G_{M2} was not detectable after tritium-labeling of the sialic acid residues in the gangliosides. There were changes detected in the labeling pattern (galactose oxidase and [^{3}H]-sodium borohydride) of gangliosides but not of the neutral glycolipids in the membranes of KB cells before or after interferon treatment. Only one small difference was noted in the labeling pattern of the glycoproteins of KB cell membrane before or after interferon treatment.

These results suggested that the initiation of the antiviral action of interferon was a membrane-related event; the initial binding of interferon to a ganglioside or ganglioside-like receptor (11, 15) seemed necessary but not sufficient to establish the antiviral state. Immediately following the interaction of interferon with its receptor, there was a change in membrane state analogous to that seen when cholera toxin or TSH interacted with thyroid membranes (21); cells nonresponsive to interferon but able to bind interferon (human KB cells) did not have the same change; the failure of KB cells to respond to interferon, despite the ability of interferon to bind to the cell surface, correlated with low levels of membrane ganglioside, which were present in high levels in membranes of sensitive mouse L_y cells. A decrease in surface exposure of the oligosaccharide moities of the gangliosides G_{M1} and G_{D1b} indicated a change in the orientation of the gangliosides in the L_y cell membrane after interferon treatment.

Effect of Interferon on Intramembranous Particle Density

A change after interferon treatment in AKR,C$^-$ cell membrane was detected morphologically by freeze-fracture electron microscopy. The number of intramembranous particles on both fracture faces were found to be significantly increased after interferon treatment for 48 hours. Figure 3A shows the fracture A face (the convex intramembranous face adjacent to the cytoplasm) from control cells. The diameter of the exposed intramembranous particles range from 5 nm to 10 nm averaging 7.5 nm. A fracture A face of membranes from cells treated with interferon for 48 hours is shown as Figure 3B. Based on particle counts from a number of micrographs, the interferon-treated membrane has twice as many particles as the control.

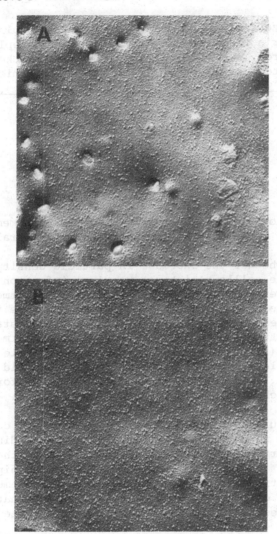

Figures 3A & 3B

Freeze-fractured replica of a portion of an AKR,C⁻ cell. The
fracture A face of (A) a control cell, (B) a cell treated with
interferon (30 units/ml) for 48 hours.

   Cells were washed five times with PBS, pH 7.4 and fixed in 2.5%
glutaraldehyde in PBS for 20 min at room temperature. After washing
three times with PBS, the cells were suspended in 30% glycerol in PBS
and droplets of the sample were put on 3 mm gold planchets for
rapid freezing first in freon, and then in liquid nitrogen. The

(Caption continued on page 92)

samples were freeze-fractured and then shadowed with platinum-carbon in a Balzar's freeze-etching apparatus. The replicas were cleaned with Chlorox, washed with distilled water, picked up on 0.25% Formvar-coated grids, and viewed in a Phillips microscope.

Temporal Relationship Between the Increase in Intramem-branous Particles and the Establishment of the Antiviral State

To determine whether the increase in particle density paralleled the increase in the cell's resistance to viral infection, both of these parameters were monitored during interferon treatment. Figure 4A shows the time course of the density increase in intra-membranous particles on exposure to interferon. The establishment of the antiviral state in AKR cells by interferon treatment followed a similar pattern (**Figure** 4B). Interferon had a notable effect on increase of particle density and inhibition of VSV yield following a 2 hour treatment. The effect became more prominent for up to 16 hours after which little if any change was observed.

Forty eight hours after the removal of interferon, cells demon-strated a decrease in antiviral activity and correspondingly, a decrease in the particle density to the same level as that in the untreated control. While a cause and effect relationship between the increase of intramembranous particles and establishment of antiviral state is difficult to demonstrate, we feel that these data strongly suggest that these two phenomena are indeed closely related.

Effect of Interferon Treatment on Plasma
Membrane Buoyant Density

Plasma membrane preparations were analysed on discontinuous sucrose density gradients. Table 1 compares the density of membranes from interferon-treated and in control cells: in the latter, 77% of the plasma membrane banded in sucrose at a density of 1.22-1.23 and 23%, at 1.23; in interferon-treated cells, 40% banded at 1.22-1.23 and 60%, at 1.23. This indicates that a pro-

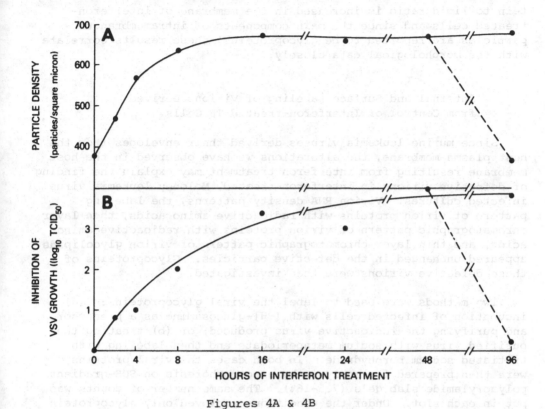

Figures 4A & 4B

Effect of varying the duration of interferon treatment on the time course of (A) increase of intramembranous particle density and, (B) establishment of antiviral activity. Broken lines represent the samples after interferon removal.

Table 1

Effect of interferon on the distribution of AKR,C⁻ cell membrane into the two density fractions

	% A_{280} in	
	Low density	High density
	$\rho \sim 1.216/1.231$	$\rho \sim 1.231$
Control	77	23
100 unit interferon/ml	40	60

tein to lipid ratio is increased in the membrane of interferon-treated cells and since the main components of intramembranous particles are reported to be glycoproteins, these results correlate with the morphological data closely.

> Internal and Surface Labeling of Virions Derived
> from Control or Interferon-treated TB Cells

Since murine leukemia viruses derived their envelopes from the host plasma membrane, the alterations we have observed in the host membrane resulting from interferon treatment may explain the finding of defective virions in interferon-treated, Moloney leukemia virus infected cultures. Virion RNA density patterns, the labeling pattern of virion proteins with radioactive amino acids, then layer chromatographic pattern of virion proteins with radioactive amino acids, and thin layer chromatographic pattern of virion glycolipids appeared unchanged in the defective particles. Glycoproteins of these defective virions were then investigated.

Two methods were used to label the viral glycoproteins: (a) incubation of infected cells with [^{3}H]-glucosamine as a precursor and purifying the radioactive virus produced; or (b) treating the purified virus with sodium metaperiodate and then labeling with tritiated sodium borohydride. In both cases the viral proteins were then prepared for analysis by electrophoresis on SDS-gradient polyacrylamide slab gels (7.5-16%). The same number of counts was put in each slot. Under the conditions employed only glycoproteins were radioactive.

In the virus labeled with [^{3}H]-glucosamine there was present a prominent band of glycoprotein with a molecular weight about 85,000 in virus derived from interferon-treated TB cells (Figure 5A) while only a trace amount of this glycoprotein was present in virus derived from control cells not treated with interferon (Figure 5B). In the case of virus tritiated after periodate treatment, a prominen glycoprotein component with a molecular weight of 85,000 (gp 85) was also present in virus from interferon-treated cells (Figure 6A). Again this glycoprotein was present only in trace amounts in virus produced by the control cells (Figure 6B).

In murine leukemia virus-infected cells, by immuno-precipitation technique, a glycoprotein around 80-90,000 in molecular weight has been shown to be the precursor of a major virion envelope glyco-protein 69/71. If indeed, the origin of the excess gp85 in virus derived from interferon-treated cells is an uncleaved precursor to gp 69/71, one effect of interferon treatment in this system must be to inhibit cleavage processing of a viral glycoprotein. MMuLV-infected TB cells treated with interferon (200 units/ml) for

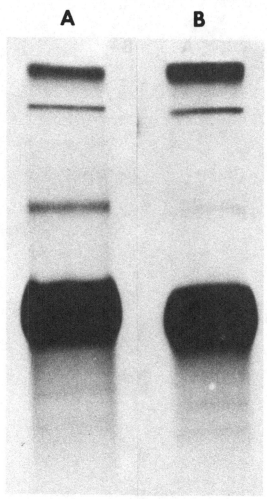

Figure 5A & 5B

Analysis of viral glycoproteins of purified MMuLV derived from
interferon-treated and control TB cells.

24 hours were labeled with [^{3}H]-glucosamine for 12-16 hours. Extracellular virions from control (B) or interferon-treated cultures (A)
were concentrated, purified, and analyzed by electrophoresis in a
7.5-16% Tris-buffered SDS-gradient polyacrylamide slab gel. The
gel was PPO-impregnated, dried, and exposed to X-OMAT R film. Phos-
phorylase A (95-100K) and bovine serum albumin (68K) were used as
molecular marker proteins. The arrow indicates the mobility of a
glycoprotein of apparent molecular weight of approximately 85,000
dalton.

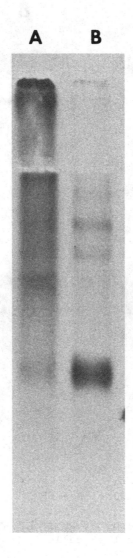

Figure 6A & 6B

SDS-gradient PAGE patterns of surface-labeled viral glycoproteins of
MMuLV derived from interferon-treated and control cells.

Purified virions sequentially treated with sodium metaperiodate and
[^{3}H]-sodium borohydride were analyzed as described in Figure 5.
(A) - interferon-treated, (B) - control.

An alteration in such a fundamental process as cleavage of viral proteins might account for the unusual inhibitory mechanism of interferon action on the production of muring leukemia viruses. The partial impairment of cleavage in turn may be due to interferon-induced alterations in the chemical (11), physical (14), and morphological (14) characteristics of the plasma membrane which have been demonstrated to take place in interferon-treated cells.

REFERENCES

1. Friedman, R.M. 1967. Interferon binding: the first step in establishment of antiviral activity. Science 156:1760-1761.

2. Berman, B. and J. Vilcek. 1974. Cellular binding character-istics of human interferon. Virology 56:646-651.

3. Besancon, F., H. Ankel and S. Basu. 1976. Specificity and reversibility of interferon ganglioside interaction. Nature 259:576-578.

4. Vengris, V.E., Reynolds, F.H. Jr., and M.D. Hollenberg et al. 1976. Interferon action: role of membrane gangliosides. Virology 72:486-493.

5. Pitha, P.A., W.P. Rowe, and M.N. Oxman. 1976. Effect of interferon on exogenous, endogenous and chronic murine leukemia virus infection. Virology 70:324-338.

6. Friedman, R.M. The antiviral activity of Interferons. Bacteriological Reviews 41, in press.

7. Besancon, F. and H. Ankel. 1974. Binding of interferon to gangliosides. Nature 252:478-480.

8. Vengris, V.E., B.D. Stollar, and P.M. Pitha. 1975. Interferon extranalization by producing cell before induction of anti-viral state. Virology 65:410-417.

9. Friedman, R.M. and L.D. Kohn. 1976. Cholera toxin inhibits interferon action. Biochem. Biophys. Res. Commun. 70:1078-1083.

10. Kohn, L.D., R.M. Friedman, and J.M. Holmes et al. 1976. Use of thyrotropin and cholera toxin to probe the mechanism by which interferon initiates its antiviral activity. Proc. Natl. Acad. Sci. USA 73:3695-3699.

11. Kohn, L.D., S.M. Aloj, and R.F. Friedman et al. Membrane glycoproteins and their relationship to the structure and function of cell surface receptors for glycoprotein hormones, bacterial toxins, and interferon. In R.E. Harmon. ed. Advances in Carbohydrate Chemistry. Centenial Meeting of the American Chemical Society, in press.

12. Lindahl, P., P. Leary, and J. Gresser. 1973. Enhancement by interferon of the specific cytotoxicity of sensitized lymphocytes. Proc. Nat. Acad. Sci. USA 69:721-725.

13. Knight, E. Jr. and B.D. Korant. 1974. A surface alteration
     in mouse L-cells induced by interferon. Biochem. Biophys.
     Res. Commun. 74:707-713.

14. Chang, E.H., J.T. Jay and R.M. Friedman. Physical and
     morphological alterations in the membrane of AKR cells follow-
     ing interferon treatment and their correlation with establish-
     ment of the antiviral state. submitted to Cell.

15. Grollman, E.F., G. Lee and R.M. Friedman et al. Interferon
     induced membrane changes: An obligitory but insufficient
     event in establishment of the antiviral state. J. Biol. Chem.
     in press.

16. Wong, P.K.Y., P.H. Yuen and R. MacLeod et al. 1977. The
     effect of interferon on de novo infection of Moloney murine
     leukemia virus. Cell 10:245-252.

17. Chang, E.H., M.W. Myers, P.K.Y. Wong et al. 1977. The
     inhibitory effect of interferon on a temperature-sensitive
     mutant cf Moloney murine leukemia virus. Virology 77:625-
     636.

18. Chang, E.H. and R.F. Friedman. 1977. A large glycoprotein
     of Moloney leukemia virus derived from interferon-treated
     cells. Biochem. Biophys. Res. Commun. in press.

19. Witte, O.N., A. Tsukamoto-Adey and I.L. Weissman. 1977.
     Cellular maturation of oncornavirus glycoproteins: topo-
     logical arrangement of precursor and product forms in cellular
     membranes. Virology 76:539-553.

20. Naso, R.B., L.J. Arcement and W.L. Karshin et al. 1976. A
     sucrose-deficient glycoprotein precursor to Rauscher leukemia
     virus gs 69/71. Proc. Natl. Acad. Sci. USA 73:2326-2330.

21. Mullin, B.R., P.H. Fishman and G. Lee et al. 1976. Thyro-
     tropin-ganglioside interactions and their relationship to
     structure and function of thyrotropin receptors. Proc. Natl.
     Acad. Sci. USA 73:1679-1683.

SELECTION OF NEW HUMAN FORESKIN FIBROBLAST CELL STRAINS FOR INTERFERON PRODUCTION

J. Vilcek, M.D.[1], E.A. Havell, Ph.D.[1], M.L. Gradoville, M.S.[1], M. Mika-Johnson, M.S.[2], W.H.J. Douglas, Ph.D.[2]

[1]Department of Microbiology, New York University School of Medicine New York, N.Y. 10016
[2]W. Alton Jones Cell Science Center Lake Placid, N.Y. 12946

ABSTRACT

The aim of this work has been to isolate and characterize new diploid cell strains, suitable for large-scale production of human fibroblast interferon. Twenty cell strains were isolated from individual neonatal foreskins obtained with the informed consent of the donors' parents. The techniques employed for the isolation of the cell strains were aimed at obtaining the highest possible yield of normal diploid cells, free of contaminating microorganism and viruses. The bulk of the cell yield has been frozen at a low population doubling level. Each of the isolated cultures was tested for interferon producing characteristics with poly(I)·poly(C) under a number of different conditions including "superinduction" with metabolic inhibitors. Most of the newly established cell strains produced lower interferon yields than the reference FS-4 cell strain. However, some new cell strains produced similar interferon yields as the FS-4 cells on superinduction.   Five cell strains, designated FS-30, FS-35, FS-44, FS-48 and FS-49, identified as the highest interferon producers among the new cells, were selected for further testing. Of these, three cell strains (FS-35, FS-48 and FS-49) produced similar interferon yields as FS-4 cells after superinduction. Cell strains FS-48 and FS-49 were found to have stable interferon producing characteristics over a wide span of population doubling levels. The interferon produced in these new cell strains had the antigenic and biological characteristics of human fibroblast interferon.

INTRODUCTION

Primary cultures of human buffy coat cells have served as the major source of interferon for clinical trials (1-5). There is a need for alternate sources of human interferon for at least two reasons: (a)The number of buffy coats that can be procured for interferon production is not unlimited. Although substantial quantities of leukocyte interferon can be obtained from this source and it may be possible to increase further the present production capacity, it seems doubtful that this system can be utilized for mass production on an industrial scale. The major limitation is that buffy coat cells cannot be serially propagated. (b)It is known that the major component of leukocyte interferon preparations is different in its antigenic and physicochemical (6-8) characteristics from the major component of interferon preparations produced in cells of nonlymphoid origin, i.e., "fibroblast interferon." Since leukocyte and fibroblast interferons also differ somewhat in their biological activities (8-10) the antiviral and antitumor efficacies of both, the human leukocyte and human fibroblast interferons, should be evaluated clinically.

The rationale for the choice of human diploid cell strains as the safest source of interferon for clinical trials has already been explained (11-12). The highest yields of interferon from diploid cell strains have been obtained after stimulation with the double-stranded polynucleotide, polyinosinic-polycytidylic acid [poly(I)·poly(C)], in combination with judicious treatment with metabolic inhibitors, termed "superinduction" (13,14).

In 1972, our laboratory started comparing interferon producing characteristics of cell strains derived from individual neonate foreskins. Among the very first cell strains established was the FS-4 line which has become widely used for studies concerned with interferon production and its regulation. Although we subsequently established many foreskin cell strains from other donors, none was quite comparable in the amounts of interferon produced to the FS-4 cells (15, and unpublished data).

There is a need for a well characterized diploid cell strain, capable of producing high yields of interferon. The FS-4 strain is, and will continue to be, available for many types of smaller-scale laboratory studies. However, the production of interferon for clinical trials will require large quantities of cells which (a)would have optimal interferon producing characteristics, and (b)would satisfy the requirements of the Food and Drug Administration as a safe source of interferon for clinical studies. In this paper we report on the isolation of new diploid cell strains from neonate foreskins and the determination of their interferon producing characteristics.

MATERIALS AND METHODS

Establishment of Foreskin Cell Strains

Newborn foreskins were obtained with informed consent of
the donors' parents. Freshly excised tissue was washed in saline
and immersed in tissue culture medium [Ham's F12K (16) supplemented
with 10% fetal bovine serum, penicillin G (100 units/ml), kanamycin
(100 µg/ml) and amphotericin B (6.25 µg/ml)]. At the cell culture
laboratory, the tissue was washed twice in fresh medium and gently
minced into 1 mm^3 fragments with sterile scissors. This mince was
distributed to at least five 150 mm Petri dishes (Falcon Intra-
grid, Fisher Scientific, Rochester, NY, 32 squares/dish). A
single foreskin fragment was placed in the center of each grid
square of the dish and the vessel incubated without medium for
60 minutes in a humidified atmosphere of 5% CO_2 in air at 37°C.
During this time, the majority of the explants attached firmly
to the dish. Thirty mls of tissue culture medium was added to
each container. Care was taken to add medium slowly to avoid
dislodging the foreskin fragments. The cultures were then returned to the CO_2 incubator and left undisturbed for one week.

Cultures were examined by phase-contrast optics to assess
cellular outgrowth from the fragments. A halo of epitheloid cells
generally surrounded the fragments at this time. The cultures
were fed every other day for the next two weeks. During this
time period, fibroblastoid cells migrated from the explant.
At the end of the third week of in vitro cultivation, the
fibroblastoid cells formed a semi-confluent culture. These
cells were designated as population doubling level zero.

The fibroblastoid cells were harvested from the Petri dishes
by enzymatic treatment. Monolayer cultures were washed twice
in saline and then incubated at 37°C in 0.1% trypsin (Difco
1:250, Detroit, MI), 0.1% collagenase (Worthington Biochemical
Corp., Freehold, NJ, CLS grade) and 1% chicken serum in saline.
Following a thirty minute treatment, the majority of the cells
rounded up and detached from the culture surface. These cells
were collected, washed twice in medium, and 5 x 10^5 viable cells
were transferred to a T-25 flask (Corning Plasticware, Corning, NY)
containing 5 ml of medium. Five days later the flask was semi-
confluent and this cell population was designated as population
doubling level one. These cells were enzymatically harvested
and 1 x 10^6 viable cells were seeded into a T-75 flask. Five
days later this flask was semi-confluent and contained 4 x 10^6
viable cells; indicating that two additional population doublings
had occurred. These cells were enzymatically harvested and
1 x 10^6 viable cells were inoculated into each of four T-75 flasks.
Cells in three of these flasks were fed with antibiotic-free medium

while antibiotics remained in the medium of only one flask. Five
days later these flasks were semi-confluent so that approximately
two additional population doublings had occurred.

These cells were enzymatically harvested and the cell
suspension adjusted to contain 4×10^6 viable cells per milliliter of medium. An equal volume of nutrient medium containing
10% dimethylsulfoxide was added dropwise. Ampoules containing
1 ml of this suspension were frozen in the following manner using
the Linde controlled rate freezer. The temperature was initially
reduced at the rate of 1°C per min down to 30°C. The cooling
rate was then increased to 5°C per min until a temperature of
-70°C was reached. At this point, the ampoules were stored in
liquid nitrogen. These ampoules contained 4×10^6 cells at the
fifth population doubling level.

Preliminary Characterization of Foreskin Cells

Sterility tests for mycoplasma, bacteria and fungi were conducted on cells in log phase growth at population doubling level
5, 15, and 25. Foreskin cells were inoculated into mycoplasma
test medium according to the procedures outline by McGarrity
(17). Cells and medium from these cultures were also tested on
fluid thioglycollate medium for bacteria (18) and soybean casein
digest medium for fungi (18). Karyological analysis of FS-48 and
FS-49 cell strains was performed at population doubling levels
10, 25, and 40. One hundred suitable metaphase cells were examined
from the cell strains at each time point (19).

Propagation of Cells for Screening

All human fibroblast cells derived from individual neonate
foreskins were propagated in Eagle's minimal essential medium
(MEM) supplemented with 5% heated fetal bovine serum (FBS).
The growth medium also contained the antibiotics gentamycin (50
µg/ml) and amphotericin B (2.5 µg/ml) and it was buffered with
4- (2-hydroxyethyl) - 1 - piperazine-ethanesulfonic acid (Hepes,
13 mM), N-[tris (hydroxymethyl) - methyl] glycine (Tricine, 6 mM),
and soidum bicarbonate (1.1 g/liter). The cultures were propagated
at 36°C in sealed plastic 250 ml flasks. The procedure used for
the preparation of cultures for the testing of their interferon
producing potential is summarized in Table 1.

Interferon Induction and Superinduction

Confluent 12-day-old cultures of each of the tested fibroblast

TABLE 1

Procedure for preparation of cultures for the testing of
interferon producing characteristics of foreskin cell strains

Day   Procedure

0   Cell suspension is obtained by trypsinization of confluent
         cultures grown in 250-ml plastic flasks. Cultures are
         seeded in 60-mm plastic Petri dishes, using 200,000 cells/
         dish in 5 ml MEM with 5% FBS. Freshly seeded cultures
         are incubated at 36°C in the presence of 4% CO_2.

6   Old growth medium is removed, 5 ml fresh MEM with 5%
         FBS re-added.

12   Cultures are used for interferon induction under condi-
         tions of "simple induction" [1-hr exposure to various
         concentrations of poly (I) · poly (C)] or "superinduction"
         (see Table 2). Cell counts are obtained from the same lot.

strains were induced with poly(I)·poly(C) (P-L Biochemicals,
Milwaukee, WI) alone (induction) or with poly(I)·poly(C) in
combination with certain antimetabolites (superinduction).
Induction consisted of invubating cultures with poly(I)·poly(C)
in 2 ml of MEM for 1 hr. At the end of this period, the medium
with poly(I)·poly(C) was removed, the cultures were washed 4
times with Hanks' balanced salt solution and replenished with 2
ml of MEM containing 2% FBS. The cultures were then incubated at
34°C for an additional 23 hr and the culture fluids were
harvested for interferon titrations.

    An outline of the interferon superinduction schedule used
during the screening of new foreskin fibroblast strains is presented in Table 2. This method employs the antimetabolites
cycloheximide (Upjohn Co., Kalamazoo, MI) and actinomycin D
(Calbiochem, San Diego, CA) added in a sequence which was found
to be optimal for the enhancement of the interferon yields (13-15).

Interferon Assays

    Titration of interferon activities was done by the micromethod
of Armstrong (20) as modified by us (13), using the human FS-7
cell strain and vesicular stomatitis virus (VSV). The bovine
MDBK cell line was used in an otherwise similar procedure to assess
the degree of heterospecific antiviral activity of interferon

TABLE 2

Outline of interferon superinduction schedule used for screening
of new foreskin fibroblast strains

Hour from onset

0   Cultures are washed once with buffered saline;
                     induction medium consisting of poly(I)·poly(C)
                     (5-20 ug/ml) and cycloheximide (10-50 ug/ml)
                     in serum-free MEM is added

5   Actinomycin D (final conc. 1-5 ug/ml) is added
                     in a small volume of MEM directly to the in-
                     duction medium on the cultures

6   Induction medium is removed and discarded; the
                     cultures are washed 4 times with saline; pro-
                     duction medium is added, consisting of MEM
                     with 0.2% (final conc.) human plasma protein
                     fraction, U.S.P. or 2% FBS

30   Production medium containing crude interferon is
                     collected

Notes: All volumes of media are 2 ml/culture grown in 60-mm
plastic Petri dishes. Incubation temperature from the time of
induction is 34°C unless otherwise indicated. Two cultures are
used per experimental group in all experiments.

preparations. Interferon titers are expressed in reference units,
corrected to the human leukocyte interferon standard G-023-901-527
(obtained from the National Institute of Allergy and Infectious
Diseases, Bethesda, MD). The interferon titers obtained in the
bovine MDBK cells were not corrected.

Interferon Neutralization Tests

     The availability of potent antiserum preparations against
either human fibroblast interferon (anti-F) or human leukocyte
interferon (anti-Le) enabled the antigenic characterization by
means of quantitative neutralization tests of interferon activity.
The two antisera were capable of neutralizing the antiviral
activity of the human interferon species which served as the
immunogen, however, neither of the two antisera was capable of
neutralizing the antiviral activity of the heterologous human

interferon species.   The anti-F interferon serum was prepared by
the immunization of an adult female New Zealand white rabbit with
purified (specific activity $>1 \times 10^7$ units/mg of protein) FS-4
interferon.   The anti-Le inerrferon serum was prepared by the
immunization of a sheep against Sendai virus-induced leukocyte
interferon and was the kind gift of Dr. C.B. Anfisen.

The quantitative neutralization test of interferon activity
was done by mixing serial two-fold dilutions of each antiserum
with an equal volume of tested interferon (final concentration
10 units/ml).   The samples were then mixed, incubated at 37°C
for 1 hr and 0.2 ml of the mixtures was then added to individual
wells of MicroTest II plastic trays containing a confluent sheet
of FS-7 cells.   The cultures were then incubated for 18 hr at
37°C in 5% CO_2.   At the end of this incubation period, the fluids
were aspirated and the cell sheets received medium containing about
1,000 PFU of VSV.   The test plates were reincubated for an additional 48 hrs at which time the cytopathic effect was scored.
The neutralizing titer is defined as the reciprocal of the highest
dilution of the antiserum that prevents the antiviral activity of
10 interferon units/ml (final concentration).

RESULTS

Screening of Interferon Producing Characteristis of
20 Newly Established Cell Lines

Groups of cultures of 3-4 new cell strains were tested side-
by-side with simultaneously seeded FS-4 cultures.   Every new cell
strain was tested in at least two independent experiments; in each
experiment interferon production was compared under five different
experimental conditions.

The experimental conditions used for testing varied somewhat
from experiment to experiment.   In each experiment one or two
groups of cultures from each cell strain were induced with
poly(I)·poly(C) under conditions of "simple induction" (see
Table 1), using the inducer at 0.1, 1 or 10-20 μg/ml.   The
remaining 3-4 groups of cultures were tested under "superinducing
conditions", varying the concentration of the components in the
induction medium in the individual groups (see Table 2).

An example of the results of such a comparison between FS-4
cells and three new cell strains is shown in Table 3.   The
results are expressed in terms of reference units produced/10^6
cells.   (Although expressing the results in this way is necessary
in order to be able to average values from different experiments,
it has the disadvantage of "boosting" the yields of those cell
lines which had lower cell counts on the day of induction.   In

TABLE 3

Comparison of interferon producing characteristics of FS-4 cells and three new cell strains.

Condition of stimulation	Composition of induction medium (μg/ml)			Interferon yield (ref. units/10^6 cells)			
	Poly(I)·Poly(C)	Cyclo-heximide	Act. D	FS-4	FS-47	FS-50	FS-51
Induction[a]	0.1	0	0	27	<7	<8	<8
	20	0	0	3,000	370	230	350
Super-induction[b]	20	10	1	52,000	9,000	19,000	19,000
	20	10	2	62,000	16,000	28,000	37,000
	20	10	5	62,000	23,000	34,000	56,000
Cell Counts/dish (x10^6)[c]				1.25	1.3	1.05	1.05

a Cells were exposed to the indicated concentration of poly (I)·poly (C) in 2 ml MEM for 1 hr, washed and replenished with 2 ml of production medium (MEM with 2% FBS). Culture fluids were collected after 24 hr of incubation at 34°C.

b See Table 2 for details of treatments. Production medium (2 ml MEM with 2% FBS per culture) was collected after 30 hr of incubation at 34°C. The results are average values from two in dependent experiments.

c At the time of induction; values are averages from two experiments.

the final analysis the most important parameter is the total yield per culture, not the yield per certain number of cells.)

FS-4 cells were superior to all the other strains shown in Table 3, particularly on induction under non-superinducing conditions. With superinduction the differences were smaller, but only in one group (FS-51 cells treated with 5 µg/ml of actinomycin D) was the yield similar to that produced by FS-4 cells. None of the new cells shown in Table 3 were selected for further testing.

All 20 new cell strains were analyzed side-by-side with FS-4 cells in a similar manner, but we do not intend to test the tolerance level of the readers of this article by showing the results of all of these experiments. Suffice it to make the following general conclusions that may be of some interest.

Varying the concentrations of poly(I)·poly(C) between 5-20 µg/ml or of cycloheximide between 10-50 ug/ml during the first 6 hr in the superinduction schedule (see Table 2) did not have a consistent influence on the resulting interferon yield. On the other hand, varying the concentration of antinomycin D (1, 2 or 5 ug/ml) did influence interferon production and the effect was somewhat cell-dependent. In FS-4 cells 2 or 5 µg/ml gave good results, with 2 µg/ml being somewhat superior in some experiments (see also Table 4). In most of the new cells the titers increased with increased concentrations of actinomycin D up to 5 µg/ml. (Concentrations above 5 µg/ml were not tested.)

On the basis of the primary screening, inferior cell lines were gradually eliminated. Finally, after repeated testing of some intermediate and high producing cell strains, five "finalist" cells were selected for another side-by-side comparison with the FS-4 line (Table 4). Although substantial differences were seen in the amounts of interferon produced on induction with poly(I)·poly(C) in the absence of treatment with cycloheximide and actinomycin D, much less variation was observed after superinduction. Three of the newly selected cell strains (FS-35, FS-48 and FS-49) produced between 48,000 and 96,000 ref. units/10^6 cells after superinduction. Yields from the two other new cell strains (FS-30 and FS-44) were slightly lower in some of the superinduced groups.

Comparison of Interferon Production by FS-4, FS-48 and FS-49 Cells.

It was necessary to ascertain that the characteristics of interferon production in the new cells remain constant over a wide range of population doubling levels. Table 5 shows the results of repeated testing with cells at different passage levels. (Note that the results are given in yields/ml, not in yields/10^6 cells.) In the "induced" groups, FS-4 cells

TABLE 4

Comparison of interferon producing characteristics of FS-4 cells and five selected high producing new cell strains.

Condition of stimulation	Composition of induction medium (µg/ml)			Production medium with	Interferon yield (ref. units/10^6 cells)					
	Poly (I)·poly (C)	Cyclo-heximide	Act. D		FS-4	FS-30	FS-35	FS-44	FS-48	FS-49
Induction [a]	1	0	0	2% FBS	780	210	45	560	40	80
Super-induction [b]	5	20	5	2% FBS	68,000	28,000	56,000	47,000	82,000	72,000
	20	20	2	2% FBS	102,000	35,000	48,000	61,000	80,000	51,000
	20	20	5	2% FBS	68,000	57,000	96,000	83,000	52,000	82,000
	20	20	5	0.2% HPPF	39,000	42,000	56,000	27,000	54,000	61,000
Cell counts/dish (x 10^6) [c]					1.1	0.9	0.8	0.8	1.0	1.3

a,b,c See Table 3. All results represent average values from two separate experiments.

TABLE 5

Comparison of interferon producing characteristics of FS-4, FS-48 and FS-49 at various population doubling levels.

Exp. No.	Cell strain	Pop.doubling level	No. of cells/dish $(\times 10^6)$	IF yield (ref. units/ml)		
				Induced[a]	Primed[b]	Super-induced
	FS-4	20	1.4	3,840	7,680	38,400
1	FS-48	10	1.2	384	7,680	25,600
	FS-49	13	1.7	1,536	5,120	51,200
	FS-4	24	1.8	1,280	10,240	51,200
2	FS-48	13	1.4	480	1,920	25,600
	FS-49	18	1.9	960	20,480	51,200
	FS-4	26	3.2	2,560	ND	76,800
3	FS-48	18	2.5	480	ND	51,200
	FS-49	23	3.1	960	ND	76,800
	FS-4	22	2.0	1,920	ND	25,600
4	FS-48	23	1.7	1,280	ND	25,600
	FS-49	26	2.6	640	ND	51,200
	FS-4	28	2.1	1,920	10,240	38,400
5	FS-48	27	1.8	512	2,560	12,800
	FS-49	30	1.8	1,024	5,120	25,600

a Cultures were induced by incubation with poly (I) · poly (C) (20 µg/ml) in 2 ml MEM for 1 hr. Thereafter the cells were thoroughly washed and replenished with 2 ml MEM containing 2% FBS. The culture fluids were collected for interferon assays after 24 hr incubation at 34°.

b Incubated with 100 units/ml of human fibroblast interferon for 17 hr prior to induction as in a.

c Cultures were stimulated by incubation with 2 ml of MEM containing poly (I) · poly (C) (5 µg/ml) and cycloheximide (20 µg/ml).
Actinomycin D (final conc. 5 µg/ml) was added after 5 hr. At
6 hr the cultures were washed and replenished with 2 ml MEM containing 2% FBS. Culture fluids were collected after 30 hr of
incubation at 34°.

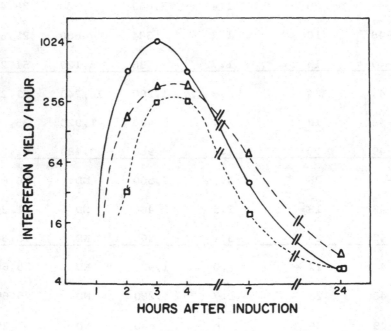

FIGURE 1

Kinetics of interferon production in FS-4 (0), FS-48 (□), and FS-49
(△) cells after stimulation with poly(I).poly(C). Cultures of
FS-4 (pop. doubling 20), FS-48 (pop. doubling 10) and FS-49 (pop.
doubling 13) cells in 60-mm Petri dishes were stimulated by exposure
to poly(I)·poly(C) (20 µg/ml) in MEM for 1 hr. The cells were then
thoroughly washed and futher incubated at 37°C with 2 ml of production medium containing 2% FBS. At the intervals indicated in the
graph, the culture fluids were collected, the cultures were washed
once with Hanks' slaine and replenished with fresh production
medium (warmed at 37°C). Interferon titers of the fluids
collected at 7 and 24 hr were divided by the number of hours which
had elapsed from the time of preceding collection.

were consistently superior to the other two lines. Priming
with interferon increased interferon yields in all three cell
strains between about 2- and 20-fold. The highest yields were
obtained from superinduced cultures, with the yields from
FS-4 and FS-49 cells somewhat superior to FS-48 cells. It is
noteworthy that FS-4 and FS-49 cells also grew to higher densities than FS-48 cells. The comparative testing is continuing
with cells at higher population doubling levels.

Kinetics of Interferon Production in FS-4, FS-48, and FS-49 Cells

It is important that the characteristics of interferon production in the newly selected cell strains resemble as closely
as possible those of FS-4 cells. It is known that a short
exposure of FS-4 cells to poly(I)·poly(C) results is a rapid
stimulation of interferon production, with the peak rate of
synthesis (measured by accumulation in the culture fluid) reached
at about 3 hr after stimulation. The kinetics of interferon production in the FS-48 and FS-49 strain were quite similar although
the peak of production may be slightly delayed as compared to
FS-4 cells (Fig. 1). The results also reflect the fact that the
amount of interferon produced in the absence of superinducing
treatments is consistently lower in FS-48 and FS-49 cells than in
the FS-4 strain.

Some Properties of Interferon Produced in FS-4, FS-48, and FS-49 Cells

It is known that interferon produced in FS-4 cells is
antigenically distinct from the bulk of interferon produced in
primary cultures of buffy coat cells (6). The results shown
in Table 6 demonstrate that interferons produced in all three
foreskin cell strains were neutralized by antibody against
fibroblast interferon but not by a monospecific antiserum against
leukocyte interferon. In contrast, preparations of leukocyte
interferon were neutralized only by homologous antibody.

Fibroblast and leukocyte interferons are known to differ
in the degree of their heterospecific antiviral activity in
bovine cells (10). Leukocyte interferon is highly active,
while fibroblast interferon exerts only a small fraction of
its homologous activity in bovine cells. In this respect too,
interferons produced in FS-48 and FS-49 cells resembled FS-4
interferon and they were clearly distinguishable from leukocyte
interferons (Table 7).

TABLE 6

Antigenic properties of interferons produced in FS-4, FS-48 and
FS-49

Source of interferon	Neutralizing titer of anti-interferon serum	
	Anti-fibroblast	Anti-leukocyte
FS-4	3,277	<102
FS-48	3,277	<51
FS-49	6,554	<26
Leukocytes (K. Cantell)	<12	2,458

TABLE 7

Heterospecific activity of interferon produced in FS-4, FS-48 and
FS-49

Source of interferon	Interferon titer on cells		Human/ Bovine
	Human (FS-7)	Bovine (MDBK)	
FS-4	8,192	64	128
FS-48	16,384	96	170
FS-49	49,152	96	512
Leukocytes (K. Cantell	49,152	98,304	0.5
Leukocytes (reference)	19,200	19,200	1

Preliminary Screening of New Cell Strains for Freedom from
Microbial Contamination and Retention of Diploid Karotype

Sterility tests for mycoplasma, bacteria, and fungi were
consistently negative on all population doubling levels tested
for the FS-35, FS-48 and FS-49 cells. The foreskin cells also
retained a diploid karyotype at all population doubling levels
tested. The following information represents the percentage of
diploid cells at each population doubling level: FS-48 cells
population doubling level 10 (93 ± 1.2%), population doubling
level 25 (94 ± 1.6%), population doubling level 40 (90 ± 2.5%);
FS-49 cells population doubling level 10 (92 ± 1.4%), population
doubling level 25 (93 ± 1.5%) and population doubling level 40
(91 ± 3.1%).

DISCUSSION

The most important requirement for a useful cellular source
of interferon is the satisfactory yield of active material.
Cells that can be serially propagated in culture are likely to be
better suited for large-scale production than primary cultures.
From the point of view of economy, an immortal cell line having
a rapid growth rate and producing high yields of interferon would
be the most suitable choice. However, current regulations pre-
clude the clinical use of biologicals produced in aneuploid cell
lines. In view of these considerations, diploid human cell strains
appear to hold the greatest promise as a practical and safe source
of interferon for human use (11, 12).

The work described in this paper has been aimed at the
selection of a diploid cell strain that would produce satisfactory
yields of human interferon and essentially meet the safety requirements devised for the production of live virus vaccines (18).
The precautions taken during the collection of the foreskin tissue
and the propagation of the master cell stock included the following:
(a) informed consent was obtained from the donors' parents, (b)
medical histories of the donors were received, (c) aseptic techniques were employed during all stages of handling of the foreskin
tissue,   (d) after the establishment of explant cultures the
cells were passaged in antibiotic-free medium, (e) all cultures
of the master cell stock were shown to be free of bacterial,
fungal and mycoplasmal contamination, and (f) preliminary karyological analysis showed that the cell strains are free of gross
karyological abnormalities.

The results included in this paper show that the interferon
yields obtainable from the highest producer cell strains (FS-35,
FS-48 and FS-49) are comparable to the yields produced by the well

characterized FS-4 strain. Further detailed studies will now be
carried out to ascertain that these cells are free from contamination with latent viruses and they fully conform to the desired
characteristics of diploid cell strains, i.e., that they lack
oncogenic potential, do not exceed the acceptable levels of chromo-
somal abnormalities, etc. It is planned that following the satisfactory completion of these studies the cells will be made available
for distribution to qualified laboratories for studies involving
interferon production.

ACKNOWLEDGMENTS

Angel Feliciano provided skilled technical assistance. This
work was supported by Contract NO1-AI-02169 from the National
Institute of Allergy and Infectious Diseases.

REFERENCES

1. Merigan, T.C., T.S. Hall, S.E. Reed, et al. 1973. Inhibition of respiratory virus infection by locally applied interferon. Lancet. 563-567.

2. Strander, H., K. Cantell, G. Carström et al. 1973. Systematic administration of potent interferon to patients with malignant diseases. J. Natl. Cancer Inst. 51:733-742.

3. Jordan, G.W., R.P. Fried, T.C. Merigan. 1974. Administration of human leukocyte interferon on herpes zoster. I. Safety, circulating antiviral activity and host responses to infection. J. Inf. Dis. 130:56-62.

4. Jones, B.R., D.J. Coster, M.G. Falcon, et al. 1976. Topical therapy of keratitis with human interferon. The Lancet ii:128.

5. Greenberg, H.B., R.B. Pollard, L.I. Lutwick, et al. 1976. Effect of human leukocyte interferon on hepatitis B virus infection in patients with chronic active hepatitis. New Eng. J. 295:517-522.

6. Havell, E.A., B. Berman, C.A. Ogburn, et al. 1975. Two antigenically distinct species of human interferon. Proc. Nat. Acad. Sci. U.S.A. 72:2185-2187.

7. Stewart, W.E. II, P. DeSomer, V.G. Edy, et al. 1975. Distinct molecular species of human interferons: requirement for stabilization and reactivation of human leukocyte and fibroblast interferons. J. Gen. Virol. 26:327-331.

8. Vilcek, J., E.A. Havell, S. Yamazaki. 1977. Antigenic, physicochemical, and biologic characterization of human interferons. Ann. N.Y. Acad. Sci. 284:703-710.

9. Edy, V.G., A. Billiau, P. DeSomer. 1976. Human fibroblast and leukocyte interferons show different dose-response curves in assay of cell protection. J. Gen. Virol. 31:251-255.

10. Gresser, I., M.T. Bandu, D. Brouty-Boye, et al. 1974. Pronounced antiviral activity of human interferon on bovine and porcine cells. Nature 251:543-545.

11. Hayflick, L. 1974. The choice of a cell substrate for preparation of human interferon. Pgs. 4-11 in C. Waymouth, ed. The production and use of interferon for the treatment and prevention of human virus infections. In Vitro Monograph No. 3 The Tissue culture Association, Rockville, MD.

12. Vilcek, J., E.A. Havell. 1975. Use of superinduction for
     the production of interferon in cultures of human diploid
     fibroblasts. In: Proc. Symp. on Clinical Use of Interferon.
     Yugoslav Academy of Sciences and Arts, Zagreb. pp. 27-33.

13. Havell, E.A., J. Vilcek. 1972. Production of high-titered
     interferon in cultures of human diploid cells. Antimicrob.
     Ag. Chemother. 476-484.

14. Billiau, A., M. Joniau, P. De Somer. 1973. Mass production
     of human interferon in diploid cells stimulated by poly
     I·C J. Gen. Virol. 19:1-8.

15. Mozes, L. W., E. A. Havell, L. M. Gradoville, et al. 1974.
     Increased interferon production in human cells irradiated with
     ultraviolet light. Infec. Immun. 10:1189-1191.

16. Kaighn, M.E. 1973. Human liver cells. In: Tissue Culture
     Methods and Applications. (Eds. P.F. Kruse, Jr. M.K.
     Patterson). Academic Press, N.Y.

17. McGarrity, G.J., L. L. Coriell. 1973. Detection of
     anaerobic mycoplasmas in cell cultures. In Vitro 9:17-18.

18. Code of Federal Regulations. Culture media for detection
     of bacteria and fungi. Title 9 Animals and Animal Products.
     Chapter 1 Animal and Plant Health Inspection Service. pp.
     11325. 1976.

19. Hsu, T.C. 1973. Karyology of cells in culture. In: Tissue
     Culture Methods and Applications. Eds. P.F. Druse, Jr., M.
     K. Patterson. Academic Press, N.Y.

20. Armstrong, J. A. 1971. Semi-micro, dye-binding assay
     for rabbit interferon. Applied Microbiol. 21:723-725.

TISSUE CULTURE MODELS OF IN VIVO INTERFERON PRODUCTION AND ACTION

F. Dianzani[1], I. Viano[2], M. Santiano[2], M. Zucca[2],
P. Gullino[3], and S. Baron[1]

[1]University of Texas Medical Branch
   Galveston, Texas 77550
[2]University of Turin
   Turin, Italy
[3]National Institutes of Health
   Bethesda, Maryland 20014

Interferon participation in recovery from viral infection is clearly documented by a series of in vitro observations. It has been shown in vitro that the inhibition of interferon production during infection often notably increases the amount of virus produced (Glasgow and Habel 1962). Cultures infected with viruses in conditions favoring good interferon production yield scarce quantities of virus and establishment of inapparent or abortive infections (Isaacs, 1963). The same effect can also be obtained by adding preformed interferon to the infected cultures.

In vivo, interferon production also depresses viral multiplication (Baron and Buckler 1963). The decline or arrest of viral production in many infections coincides with maximal interferon production and sometimes viral multiplication begins again as soon as the inhibitor disappears from circulation and tissues. These effects can also be reproduced and amplified by the administration of exogenous interferon or interferon inducers. Many conditions which tend to worsen the course of viral infections, as for example cortisone therapy, physical and emotional stress, etc., often cause a clear decrease in interferon production (Kilbourne et al. 1961; Jensen 1968). Furthermore animals with a genetically defective interferon system are clearly more vulnerable to experimental viral infections than normal animals (Hanson et al., 1963). Analogous observations were also recently made in human viral diseases, show-

ing that lymphocytes of individuals with long, serious infections
manifest a significant reduction in interferon production capacity
and this defect may be genetically determined (Emodi and Just 1974;
Tolentino et al. 1975).

It may, therefore, be concluded that the development of many
viral infections can be affected by the induction of endogenous interferon. However, several observations raise the possibility that,
in vivo, the protective effect of interferon produced during infection may be much more effective than the protective effect of applied
interferon. For example, interferon levels during mild infections
of animals or man often may be low and of short duration (Gresser
and Naficy 1964; Baron et al. 1966) whereas experimental therapy
with applied or induced interferon generally requires high levels
of interferon applied for longer periods of time (Worthington and
Baron 1971, Worthington et al. 1973). This consideration applies
especially to treatment of human infections, where it has been shown
that in order to obtain appreciable protection, a much larger quantity of interferon is needed (Merigan et al. 1973) than that antici-
pated. Until a short time ago this finding could be explained by the
rapid elimination of interferon, at least partially, by the kidney
(Bocci et al. 1967; Ho and Postic, 1967) or inactivation by body
fluids (Stanton and May, 1974). In other words, it was thought that
administered interferon did not remain in contact with the cells long
enough to induce adequate resistance. Recent findings have demonstrated, however, that human cells exposed to moderately elevated
amounts of interferon develop significant elevated antiviral resistance after as little as 1 minute of contact with the inhibitor
(Dianzani and Baron, 1975) (Fig. 1). Therefore, the requirement
for large amounts of exogenous interferon may in part be due to a
need of high local concentrations of interferon.

The biological significance of this rapid effect is clear. In
vivo, the infected cell frequently produces both virus and interferon. A substantial lead of the host over the spreading virus would
be provided by a rapid action of the interferon on surrounding cells

It is possible to estimate the possible range of interferon
concentrations in the extracellular fluid around producing cells
in vivo to determine whether these concentrations are sufficient to
induce the early appearing resistance under natural conditions.
The assumptions made for the estimation are: that single infected
cell releases its interferon into the surrounding intercellular
space over a period of six hours; that the half-life of interferon
in the extracellular fluid is similar to the rapid turnover phase
in serum, i.e., 10 min; the volume of extracellular space surrounding
a cell in a solid tissue ranges from 12 to 120 μ^3; and that the interferon production in vivo may cover the same range as in cell culture i.e., 10 to 10,000 units/ml. As listed in the estimated range
in table 1, much more than the minimum of 30 units per ml required

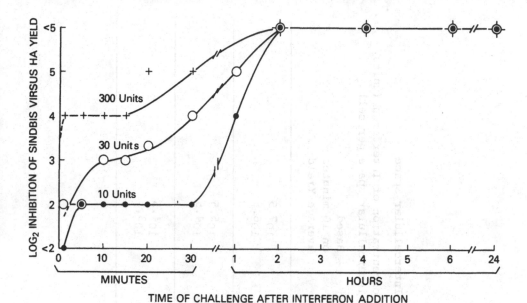

Figure 1

Effect of interferon concentration on the level of rapidly developing resistance.

to induce early resistance would be found in the extracellular fluid around the interferon producing cells in vivo.

This problem was experimentally approached using micropore chamber incorporated into the subcutaneous tissues of rats (Gullino et al. 1964). The chambers were implanted 7-10 days before the experiments and during this period of the time the host's cells proliferated around them. However, the cells were unable to pass through the membranes (0.45 μ porosity) but the fluid bathing the cells could diffuse through and be collected. The diffusion was rather slow and at least 3 hours were needed to obtain 20 μl of fluid. Immediately before the experiments the fluid in the chamber was collected (time 0) and then 0.1 ml of a suspension of Newcastle disease virus (NDV) containing 10^9 ID_{50} per ml was injected into the chamber. At preestablished times the fluid contained in the chambers was collected, diluted 1/100 and titrated for interferon activity after inactivation of residual NDV either by 7 days pH 2 treatment or by treatment with antibody to NDV. The results are reported in Table 2.

Table 1

Estimated Range of Concentrations of Interferon in the Intercellular Space

Units of Interferon produced by 2×10^5 cells Cells	Interferon yield per cell (units)	Volume of intercellular space per cell (u^3)	Estimated Concentration of Interferon (units/ml) in the intercellular space per cell	
			Based on Total Yield	Based on 10 minutes average Yield
10^4	0.05	12	10^9	$10^{7.5}$
		120	10^8	$10^{6.5}$
10^2	0.0005	12	10^7	$10^{5.5}$
		120	10^6	$10^{4.5}$
10	0.00005	12	10^6	$10^{4.5}$
		120	10^5	$10^{3.5}$

Table 2

Interferon appearance in micropore chambers implanted subcutaneously in the rat stimulated locally with NDV

Interferon titer ($\log_{10}$ units per ml) at the indicated hour after stimulation with NDV

0	1	2	3	5	24	48	72
<2	ND	5.3	ND	4.8	4.9	ND	<2.0
<2	ND	ND	5.3	ND	4.8	2.2	<2.0
<2	ND	ND	4.2	ND	4.0	<2.0	<2.0
<2	ND	ND	4.0	ND	3.5	<2.0	<2.0
<2	2.0	3.0	4.0	4.0	4.0	3.0	2.7
<2	2.0	3.5	ND	4.0	3.5	3.0	2.5
<2	<2.0	4.0	ND	3.0	4.5	3.5	2.0

It may be seen that in some samples interferon was detected
as early as 1 hour after the induction and that the titers in
the 2-5 hour samples were remarkably high and clearly exceeding
the concentration needed to induce the rapidly evolving antiviral
effect in tissue cultures. Since diffusion into the chamber
is rather slow (Swabb et al. 1974), it may be inferred that
at each of the early times the interferon detected represents
a fraction of the amount actually produced. To test this
hypothesis the diffusion rate of interferon was measured in vitro
in chambers filled with Eagle's medium supplemented with 2%
fetal calf serum and immersed in the same medium containing
3000 reference units of interferon per ml. After the indicated
times of incubation at 37° the interferon titer was determined in
duplicate chambers and the surrounding fluid. The results are
shown in figure 2.

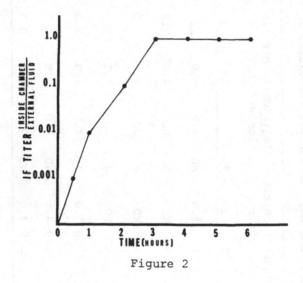

Figure 2

Kinetics of diffusion of mouse interferon into micropore chambers.

It may be seen that after 2 hours the interferon concentration
with the chambers was only 1% of the external concentration
and that the equilibrium was established between 2 and 3 hours
of incubation. This finding supports the view that in vivo
experiments large amounts of interferon were produced within
one hour after the induction with NDV.

The effects of these locally high concentrations in interferon
on the interferon-producing cell, however, is unknown. Therefore,

a study was undertaken utilizing differences in interferon concentration around producing cells to help define some of the variables which govern the action of interferon on the same cells which produce the interferon. All the experiments were performed using mouse L cells grown either in suspension (for interferon induction) with Eagle's spinner medium or in monolayers with regular Eagle's medium (for interferon assay). The media were supplemented with 10% fetal calf serum for growth and 2% for the experiments. Specifically the experiments were designed to vary the concentration of secreted, extracellular interferon and to determine the resulting kinetics of development of the resistant state. The basic manipulation was to vary the extracellular volume around a constant number of interferon producing cells such that secreted interferon would occur in different extracellular concentrations but where we would expect no alteration in the amount and concentration of the production of intracellular interferon. Mouse L cells suspended in Eagle's spinner medium were stimulated to produce IF with NDV at an input multiplicity of 50 EID_{50} per cell to simultaneously induce the entire cell population. Virus adsorption was performed at 0°. One hour later unadsorbed virus was removed by 3 cycles of centrifugation in a refrigerated centrifuge (500 g for 5 minutes) and the cells were resuspended at a concentration of 5×10^6 cells per ml in prewarmed Eagle's medium supplemented with 2% fetal calf serum and 0.025 M Herpes buffer. One ml samples of the above suspension were then distributed either in empty tubes or in centrifuge bottles containing 99 ml of prewarmed Eagle's medium. Consequently, the two kinds of vessels contained the same total number of cells but one had a cell concentration of 5×10^6 and the other had 5×10^4 cells per ml. The cells were incubated in a waterbath at 37° with gentle agitation and, at preestablished times, duplicate samples were collected. The fluids were assayed for IF and the cells were washed and challenged with mengovirus to establish the level of antiviral resistance. The mengovirus HA yield was determined after overnight incubation. Comparable sets of cells, run in parallel but not induced with NDV, were used as virus controls.

The results of a representative experiment are shown in Fig.3. It may be seen that the antiviral state developed much more rapidly in the cells incubated at high density as compared with those incubated at low density. For example, significant resistance developed at 30 minutes in the high density cell population and rapidly progressed to complete inhibition of yield of viral HA at 4 hours. In comparison, in the low density cell population, the antiviral state was first detectable at 2 hours, the rate of development of the resistance was slower and the yield of viral HA was not completely suppressed. In both these experimental conditions the development of the antiviral state was prevented by pretreatment of the cells with actinomycin D (2 μg per ml per 1 hour; Table 3), indicating that it was mediated by interferon.

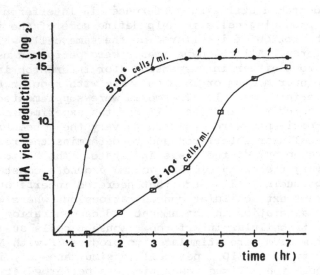

Figure 3

Development of antiviral resistance to mengovirus in cells
induced with NDV (50 ID_{50}/cell) and subsequently resuspended at
different cell concentrations. Arrows indicate inhibition of
HA yield at the lowest dilution tested.

The same cultures which were used to study the development of
resistance were used to .follow the production of interferon and its
concentration at various times after induction by NDV. The results
are shown in Table 4. At high cell concentration where the interferon was secreted into a small extracellular volume, the concentration of interferon in the medium surrounding the cells was high
(100 to 320 units per ml). In comparison, the cells at low density
(the same number of cells suspended in a large volume of medium)
produced low concentrations of interferon (2 to 4 units per ml) although the total amount of interferon produced was the same as for
the cultures at high density. Thus, the production of interferon
per cell was the same under both conditions, but at high density
the concentration of interferon around each cell was substantially
greater (Table 4) and was associated with the development of greater
antiviral activity (Figure 3). Taken together this data clearly
indicate that the degree of antiviral state attained by a cell is
related, rather than to the total amount of interferon produced, to
its concentration in the extracellular space.

The findings show in addition, that also the time of establishment of antiviral resistance is greatly influenced by the interferon

Table 3

Effect of Actinomycin D on the Development of the Antiviral State Induced by NDV in Mouse L Cells

Cell Concentration (Cells/ml)	Treatment with actinomycin D	Inhibition of Mengovirus HA yield (log_2) at the indicated (hour) after Induction by NDV				
		0.5	1	2	3	4
5×10^6	No	2	7	11	13	16
5×10^6	Yes	0	3	4	4	4
5×10^4	No	0	0	2	4	5
5×10^4	Yes	0	0	0	0	2

Table 4

Effect of Concentration of Mouse L Cells on Production of Interferon After Induction by NDV

Cell concentration (cells per ml)/ Total number of cells	Interferon concentration (units per ml)/total units of interferon produced at the indicated hour					
	0.5	1	2	3	4	7
$5 \times 10^6 / 5 \times 10^6$	<10/<10	<10/<10	<10/<10	100/100	320/320	320/320
$5 \times 10^4 / 5 \times 10^6$	<2/<200	<2/<200	<2/<200	2/200	4/400	4/400

concentration in the extracellular fluid. For instance the high
concentrations of interferon around densely packed producing cells
can be responsible for the resistance to virus challenge applied 30
minutes after induction of interferon. This appearance of resistance
to virus challenge at 30 minutes as seen in Figure 3 does not necessarily correspond to the time of the transcription and translation
of the antiviral protein(s) because there is a 30-45 minute interval
after virus challenge when newly produced AVP can still block virus
replication (Dianzani et al. 1976). Thus a conservative estimation
of the time required for production of the antiviral protein by the
concentrated cells must include the challenge time (30 minutes) plus
the time necessary for derepression of the AVP cistron (30-45 min.).
It may therefore, be inferred that the actual production of AVP occured not later than 60-75 minutes after induction by NDV.

Since interferon synthesis must precede production of AVP
(Dianzani et al. 1970) it may be concluded that under the present
conditions the time of first production of interferon must have
occured not later than 30-45 minutes. This time period may be
sufficient not only to protect the surrounding cells from the
spreading viral infection, but perhaps also influence the course
of the viral infection in the interferon-producing cell.

The dependence of resistance in the interferon-producing cell
on cell density indicates that interferon must be externalized by
the producing cell before it acts on the same cell to induce resist
ance.   This conclusion is based on the findings that: (a) the
concentration of interferon in the fluid around cells is a major
determinant of the degree of antiviral activity (Bron et al. 1967);
(b) in the present study the interferon-producing cells yielded
the same total amount of interferon per cell regardless of cell
density; and (c) the concentration of secreted interferon in the
fluid surrounding the producing cells was inversely proportional
to the cell density. Thus, the interferon-producing cells at
different densities developed different levels of antiviral activity
in relation to the concentration of interferon in the surrounding
fluid rather than in relation to the same interanl (total) amount
of interferon which they produced. This interpretation is consis-
tant with previous studies which indicated that interferon was
externalized by the producing cell before it could induce an
antiviral effect in the same cell (Lebon et al. 1975; Vengris
and Pitha 1975).

Although it does appear that interferon must be externalized
by the producing cell before inducing antiviral activity in that
cell, the degree of resistance which develops is much greater than
that which would develop in nonproducing cells exposed to the same
concentration of interferon. For example, as shown with the low
cell density population in Table 3, resistance of greater than
32,000 fold inhibition of virus yield was associated with only 4

units per ml of secreted interferon. If these 4 units/ml of interferon were applied to nonproducer cells, the final level of resist
ance attained would be only 1.5 $\log_{10}$ inhibition of virus yield
(Baron et al. 1967). This finding, taken together with the requirement for externalization of interferon for antiviral action
in the producing cells, implies that the secreted interferon may
occur in high concentration at the membrane of the producing cell
(before diffusing away into the medium) in order to induce the
strong antiviral activity. This possibility, that the site of
secretion is close to the membrane receptor for interferon
action, merits further study. Such a mechanism could also explain
the earlier findings that interferon-producing cells develop
detectable resistance to virus before they begin to produce
detectable levels of interferon (Baron et al. 1967).

In summary it seems reasonable to conclude that under
physiological conditions and in a solid tissue a cell infected
by a potent IF inducer may under centain conditions have enough
time to protect the surrounding cells (and perhaps itself)
before the production or the release of progeny virions. The
protection of cells infected with rubella virus (Wong et al. 1967)
may be such an example. The requirement for such high extracellular
interferon concentrations may also explain in part the weak
protection following systemic treatment of man with exogenous
interferon and stresses the need for a larger production of more
potent preparation of human interferon.

REFERENCES

1. Baron, S., Buckler, C.E.   Science 141:   503, 1962.
2. Baron, S., Buckler, C.E., McCloskey, R.V., Kirchstein, R.L.
    J. Immunol. 96:   12, 1966.
3. Baron, S., Buckler, C.E., Levy, H.B., Friedman, R.M. Proc. Soc. E
    Exp. Biol. Med. 125:   1320, 1967.
4. Bocci, V., Russi, M., Ritag. Experientia 23:: 309, 1967.
5. Dianzani, F., Gagnoni, S., Buckler, C.E., Baron, S. Proc. Soc.
    Exp. Biol. Med. 133:: 324, 1970.
6. Dianzani, F., Baron, S. Nature 257:   682, 1975.
7. Dianzani, F., Levy, H.B., Berg, S., Baron, S. Proc. Soc.
    Exp. Biol. Med. 152:   593, 1976.
8. Emodi, G., Just, M. Acta Paed. Scand. 63:   183, 1974.
9. Glasgow, L.A., Habel, K.J. Exp. Med. 115:   503, 1962.
10. Gresser, I., Naficy, K. Proc. Soc. Exp. Biol. Med. 117:
    285, 1964.
11. Gullino, P.M., Clark, S.H., Grantham, F.H. Cancer Res.
    24:   780, 1964.
12. Hanson, B.H., Koprowski, H., Baron, S. Microbiols 1B: 51, 1969.
13. Ho, M., Postic, B. Nature 214:   1230, 1967.
14. Kilbourne, E.D., Smart, K.M., Pokorny. Nature 190:   650, 1961.
15. Jensen, M.M. Proc. Soc. Exp. Biol. Med. 128:   174, 1968.
16. Lebon, P., Moreau, M.C., Cohen, L., Chany, C. Proc. Soc.
    Exp. Biol. Med. 149:   108, 1975.
17. Merigan, T.C., Reed, S.E., Tyrrel, D.A. Lancet 1:   563, 1973.
18. Stanton, J., May, D.C. Personal communication through
    I.S.M. A/37, June 1974.
19. Swabb, E.A., Wei, J., Gullino, P.M. Canc. Res. 34:   2814, 1974.
20. Tolentino, P., Dianzani, F., Zucca, M., Giacchino, R.J.
    Inf. Dis. 132:   459, 1975.
21. Vengris, V.E., Stollar, B.D., Pitha, P.M. Virology
    65:   410, 1975.
22. Wong, K.T., Baron, S., Ward, T.G. J. Immunol. 99:   1140,
    1967.

THERMAL AND VORTICAL STABILITY OF PURIFIED HUMAN FIBROBLAST INTERFERON

J.J. Sedmak, Ph. D., P. Jameson, Ph. D., and S.E. Grossberg, M.D.

Department of Microbiology, The Medical College of Wisconsin, Milwaukee, Wisconsin 53255 USA

ABSTRACT

The loss of biological activity upon heating or agitation of human interferons is markedly altered by changing their aqueous environment. Low pH significantly stabilizes liquid fibroblast interferon at 68°C and 37°C whereas chaotropic salts stabilize at 68°C but not at 37°C; this anomalous result may be due to reactivation of biological activity at the higher temperature. The concentration of extraneous proteins influences the apparent thermal stability at any temperature and pH; thus, interferon was not stable even at low pH at protein concentrations less than 5 µg/ml. Solutions of partially purtially purified fibroblast interferon can be inactivated by mechanical stress; the addition of proteins or nonionic detergents prevents such inactivation. Freeze-dried preparations show the greatest thermal stability. The use of high-temperature, accelerated storage tests makes it possible to predict the shelf-life of freeze-dried interferon.

INTRODUCTION

Although human leukocyte and fibroblast interferons have generally been considered to be very stable molecules, especially at low pH, it is now realized that they can be readily inactivated by a variety of physical and chemical treatments (1, 2). Leukocyte interferons are generally more stable than fibroblast interferons to both heating (3, 4, 5, 6) and mechanical stress (7, 8). Based on the stability of partially purified leukocyte interferon preparations, Mogensen and Cantell (9) suggested that addition of stabilizers is not necessary for long-term storage. However,

133

purified fibroblast interferon in dilute solutions is rapidly
inactivated at 4° (10) or under conditions of mechanical stress (8).
Its considerable potential for clinical use can only be realized
when large quantities of purified, stable interferon are available.
We have investigated the thermal and agitational (vortical) stabilit;
of highly purified, human fibroblast interferon to determine the
additives, environmental conditions or modifying procedures, alone
or in combination, that will best preserve biological activity and
ultimately provide a non-toxic material safe for administration
to human beings.

MATERIALS AND METHODS

Interferon

Poly I·poly C induced interferon was produced by J. Vilcek (11)
and partially purified by C. Anfinsen (12). The preparation
contained 500 µg/ml of added cytochrome C and had a specific activit;
of 1.6×10^6. The stability tests, unless otherwise specified,
used interferon diluted to 25 µg/ml protein.

Interferon Bioassay

Interferon potency was determined with a hemagglutinin-yield-
reduction assay using encephalomyocarditis virus as the challenge
virus in the BUD-8 strain of human skin cells (13).

Thermal Inactivation

Thermal stability of 1.0 ml liquid preparations was determined
either in accelerated isothermal test at 68°C or by long-term storag∈
tests at -70°C, -20°C, 4°C, 20°C and 37°C. Thermal stability of
freeze-dried preparations was determined in a linear nonisothermal
accelerated storage test (14) in which the temperature was increased
at 0.0125°/minute. All samples were stored at -70°C until tested
for interferon potency.

Vortes Agitation (vortical) Inactivation

This test consists of vigorous agitation of 1.0 ml of interferon
preparations at full speed with a Lab-line Instruments vortex action
mixer. The agitation was done in 30-second bursts at 4°C to prevent
thermal inactivation as a result of heat gain due to friction.

Freeze-drying

One ml of interferon was freeze-dried in a Virtis freeze-drier

model 10-700 without controlling the gaseous atmosphere above the
samples. The final residual moisture was not determined.

Chemicals

Ovalbumin, twice crystallized, was purchased from Aldrich
Chemical Co., Inc., Milwaukee, WI. Bovine plasma albumin (5 x
crystalline) was obtained from Metrix, Chicago, IL. Horse heart
cytochrome C (90-100% pure) was obtained from Grand Island Biological Co., Grand Island, NY. Gelatin was purchased from Difco
Laboratories, Detroit, MI. The tripeptides, glycylglycylglycine
and glycylglycylphenylalanine, were obtained from Sigma Chemical
Co., St. Louis, MO. Antifoam B was obtained from Dow Corning
Corp., Midland, MI. Sodium dodecyl sulfate (SDS) was obtained
from Matheson, Coleman and Bell, Norwood, OH. Tween 80 and Tween
20 was purchased from E.H. Sargent & Co., Chicago, IL. Thioctic
acid was obtained from Fluka A.G., Buchs, Switzerland. All other
chemicals were reagent grade or better.

RESULTS

Thermal Stability

Based on our findings of stabilization of mouse fibroblast
interferon to heating by chaotropic salts (15), i.e., salts whose
anions increase the solubility of nonpolar regions of proteins,
we investigated the effects of chaotropic salts on human fibroblast interferon. As seen in Figure 1, 2 M guanidine hydrochloride
(GuHCl) was even more effective than low pH in protecting the human
fibroblast interferon from thermal inactivation in our high-temperature screening test at 68°C. Our long-term stability tests at the
lower temperatures under which this interferon might be stored or
handled (e.g. 4°C or 20°C) showed that GuSCN did not provide the
expected degree of stabilization (Figure 2). The results of long-
term testing, in progress for almost one year, have indicated that
all preparations were quite stable at -70°C, as expected.

After long-term storage at -20°C the pH 7 control and the pH 2
and GuSCN preparations maintained most of their activity. The
NaSCN preparation lost considerable activity, possibly because the
high salt altered the eutectic point such that the preparation remained incompletely frozen.

At 4°C the pH 2 preparation retained 95% of its activity even
after 46 weeks of storage. At this temperature the pH 7 sample
maintained greater than 30% of its activity at 12 weeks, but by 46
weeks its activity had dropped to 7% of the starting activity. Although the NaSCN preparation lost >98% of its antiviral potency in

3 weeks, it did remain fairly stable at 4°C for 7 days, a period
which might suffice for a number of preparative or purification
procedures. Unexpectedly, the GuSCN preparation, which by our
high temperature tests should have been the most stable, lost 85% of
its activity in 1 week at 4°C. The reduced protection by GuSCN and
NaSCN at 4°C might be related to their lower solubility at this
temperature.

The stabilities seen upon storage at 20°C were similar to those
observed at 4°C. Again, the pH 2 preparation was the most stable,
retaining approximately 50% of its activity after 2 weeks; whereas
the pH 7 sample lost about 90% of its activity in 1 week. At 20°C
NaSCN provided some protection relative to the pH 7 sample over a
period of 2 weeks, suggesting NaSCN might be adequate for purification procedures, such as during elution of interferon from hydrophobic ion-exchange or immunoaffinity columns at room temperature and
brief storage therafter. The GuSCN preparation was inactivated more
rapidly than the pH 7 sample; it lost 80% of its activity in 1 day
and retained only 3% residual activity after 2 days. The anomalous
protection observed at the high screening temperature with much
more limited protection at lower temperatures may be related to the
reactivation of interferon at the high temperature. (Jariwalla,
unpublished observation).

In view of the anomalous results observed with GuSCN and NaSCN
in high-temperature tests, screening tests were undertaken at lower
temperatures with the expectation they might be more predictive.
The more physiological temperature 37°C was therefore used for
screening: there was sufficient loss in activity of the control
at pH 7 to allow stability experiments to run conveniently over
three to four days.

Some agents that failed to stabilize the purified interferon
in the 68°C screening test had some stabilizing effect in the 37°C
test. As seen in Table 1, Experiments 1 and 2, pH 2 which stabilize(
at 68°C also stabilized at 37°C; in addition, 1% bovine plasma
albumin, 10% glycerol and 0.01% Tween 80, although not nearly as
effective as pH 2, provide some stability to the interferon preparation. NaSCN and GuSCN did not provide protection in this low
temperature screening test (Table 1, Expt. 3). Experiments 2
through 7 indicate that 10% sucrose, 5% trichloroacetic acid, 5%
perchloric acid, 1% polyvinylpyrrolidone, 0.1 M $MgCl_2$, arginine,
10% polyethylene glycol and the tripeptides glycylglycylglycine and
glycylglycylphenylalanine failed to enhance the stability of the
liquid interferon preparation. Unexpectedly, the combination of
0.001% Tween 80 in the pH 2 buffer failed to stabilize the interferon (Expt. 5). The possibility that the Tween 80 prevented the
hydrogen ions from unfolding and stabilizing the interferon was
investigated by preparing the interferon in the pH 2 buffer first
before adding the Tween 80; the results of this sequential treat-

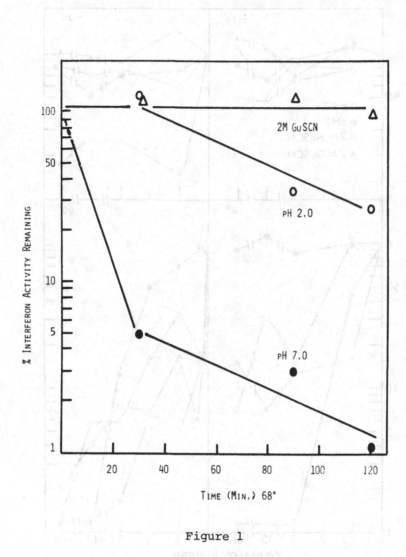

Figure 1

Stabilities of liquid preparations of purified human fibroblast interferon to heating at 68°C. One ml containing 25 µg protein was heated in the following solutions: 0.1 M sodium phosphate pH 7, ○; 2 M guanidine thiocyanate in 0.1 M sodium phosphate pH 7, △; 0.1 M KCl-HCl pH 2, ●. Samples were taken at the indicated times, cooled to room temperature, then frozen at -70° until assayed for interferon potency.

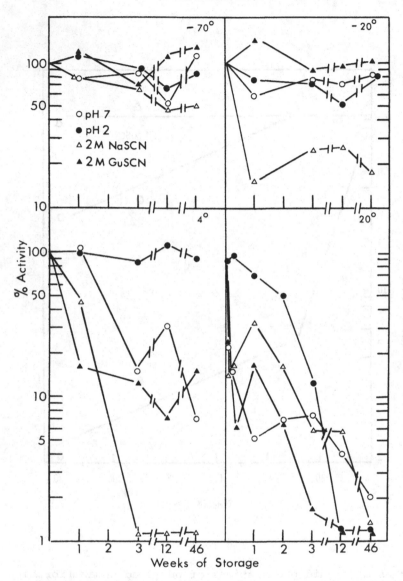

Figure 2

Long term multiple isothermal stability testing of liquid preparations of purified human fibroblast interferon. One ml aliquots
contained 25 μg protein in the following solutions: 0.1 M sodium
phosphate, pH 7, ○; 2 M sodium thiocyanate in 0.1 M sodium phosphate, pH 7, △ ; 2 M guanidine thiocyanate in 0.1 M sodium phosphate,
pH 7, ▲ ; and 0.1 M KCl-HCl pH 2, ● were stored at -70°C (upper left),
-20°C (upper right), 4°C (lower left) and 20°C (lower right for the
indicated times. Samples were taken and stored at -70°C until
assayed for interferon potency.

Table 1

Evaluation of Various Conditions on Thermal Stability of immuno-affinity-purified human fibroblast interferon in a screening test at 37°

Exp.	Condition	Original Titer	Titer after time of storage			
			3 hrs	1 day	3 days	4 days
1	pH 7,0.1 M Na-phosphate	2,900	1,150	360		140
	pH 2,0.1 M KCl-HCl	4,000	3,000	2,400		3,150
	pH 7,1% BPA	3,500	2,900	2,900		1,150
2	Saline (PBS) 0.01 M Na-phosphate	600		185	20	
	10% Sucrose, PBS	1,100		55	<20	
	10% Glycerol, PBS	3,100		1,100	600	
	0.01% Tween-80, PBS	1,800			600	
3	0.1 M Na-phosphate, pH 7	3,100		55		<50
	2 M NaSCN, 0.1 M Na-phosphate pH 7	3,800		180		55
	2 M GuSCN, 0.1 M Na-phosphate pH 7	3,800		110		<50
	5% TCA precipitate, 0.1 M Na-phosphate, pH 7	2,800		880		232
	5% PCA precipitate, 0.1 M Na-phosphate, pH 7	3,600		720		<50
4	0.1 M Na-phosphate, pH 7.0	3,200		350	<30	
	0.05 M Tris-HCL, pH 7	3,600		250	70	
	0.01% Tween-80 Tris-HCl, pH 7	4,500		1,200	1,000	
	1% Polyvinylpyrrolidone Tris, pH 7	1,200		100	70	
	0.1 M MgCl$_2$-Tris, pH 7	2,500		82	60	

Table 1 (continued)

Exp.	Condition	Original Titer	3 hrs	1 day	3 days	4 days	7 days
	0.1 M CaCl$_2$-Tris, pH 7	2,000		35	<30		
	100 µg/ml Arginine-Tris pH 7	1,600		110	85		
5	pH 7, 0.1 M Na-phosphate	4,200		220	140		
	pH 7, 0.001% Tween-80	4,000		800	1,400		
	pH 7, 0.5 mg/ml GlyGly-Glycine	3,600		950	105		
	pH 7, 1 mg/ml GlyGly-Glycine	4,600		200	135		
	pH 7, 5 mg/ml GlyGly-Glycine	3,600		1,450	580		
	pH 7, 10 mg/ml GlyGly-Glycine	3,600		950	135		
	pH 2, 0.1 M KCl-HCl	4,000		4,000	4,000		
	pH 2, 0.1 M KCl-HCl 0.001% Tween-80	4,000		450	120		
6	pH 7, 0.1 M Na-phosphate	5,000		360	170		
	pH 7, 2.0 mg/ml GlyGly-Phe	5,000		360	100		
	pH 2, 0.1 M KCl-HCl	6,000		7,800	4,500		
7	0.1 M Na-phosphate, pH 7	6,300		1,200		120	50
	0.1 M Na-phosphate, pH 7 with 10% Polyethylene glycol	13,000		1,500		120	50

Table 1 (continued)

Exp.	Condition	Original Titer	Titer After Time of Storage				
			3 hrs	1 day	3 days	4 days	7 days
	0.1 M Na-phosphate, pH 7 with 0.0001% Tween-80	9,000		1,500		800	800
	0.1 M KCl-HCl, pH 2	12,500		10,000		9,000	4,900
	0.1 M KCl-HCl, pH 2 with 0.001% Tween-80	12,500		700		125	65
	0.1 M KCl-HCl, pH 2 to which Tween-80 added to 0.001% after interferon	12,500		1,200		270	105

ment shown in Experiment 7 indicate that the interferon was less active after 4 days at 37°C than the pH 2 control sample. Possibly the Tween 80 in combination with the low pH unfolds the interferon molecule to such an extent that it cannot refold to the active form when the solution is neutralized to pH 7 for cell culture assay, as the molecule might be able to do when the pH 2 buffer is used alone during the 37°C screening test.

Since pH 2 seemed to be the best condition tested for storing the liquid interferon preparations, other pH values were tested to determine the limits of the pH range for stabilization. As seen in Table 2, only environments at pH 2 and 3 provided protection, with the interferon being rapidly inactivated in 1 day at pH values 4, 5, 6, 7, and 8. The addition of heavy water (D_2O) to a 50% concentration did not affect the stability of the interferons at pH values 2, 6 and 7 but may have enhanced stability at pH 4. However, the stability observed at pH 4 with D_2O was not comparable to that provided by either pH 2 or pH 2 with D_2O.

It should be noted that the interferon preparation tested above at 37°C contains 25 µg/ml of protein. We examined the effect of varying the protein concentration on the stability of the immuno-affinity purified interferon (Figure 3). At a protein concentration of 500 µg/ml the interferon was significantly more stable at pH 7 than it was at 50 or 5 µg/ml. The interferon was more stable at

Table 2

Effect of pH and deuterium oxide on the 37° stability of
immunoaffinity-purified human fibroblast interferon

Condition	Original Titer	Titer After Indicated days of storage at 37° (% residual activity)	
		1	2
pH 2 (0.1 M KCl-HCl)	12,000 (100%)	6,000	5,500
pH 3 (0.1 M Citric acid-Na_2HPO_4)	12,000 (100%)	9,000 (75)	5,000 (45.5)
pH 4 (0.1 M Citric acid-Na_2HPO_4)	9,500 (100%)	750 (8)	450 (5)
pH 5 (0.1 M Citric acid-Na_2HPO_4)	6,000 (100%)	850 (14)	610 (10)
pH 6 (0.1 M Na_2HPO_4-NaH_2PO_4)	6,000 (100%)	850 (14)	450 (8)
pH 7 (0.1 M Na_2HPO_4-NaH_2PO_4)	2,400 (100%)	300 (12)	110 (5)
pH 8 (0.1 M Na_2HPO_4-NaH_2PO_4)	4,000 (100%)	220 (6)	110 (3)
pH 2 with 50% D_2O	9,000 (100%)	6,000 (70)	6,000 (70)
pH 4 with 50% D_2O	6,000 (100%)	3,900 (65)	950 (16)
pH 6 with 50% D_2O	2,400 (100%)	220 (9)	400 (18)
pH 7 with 50% D_2O	1,900 (100%)	220 (12)	50 (3)

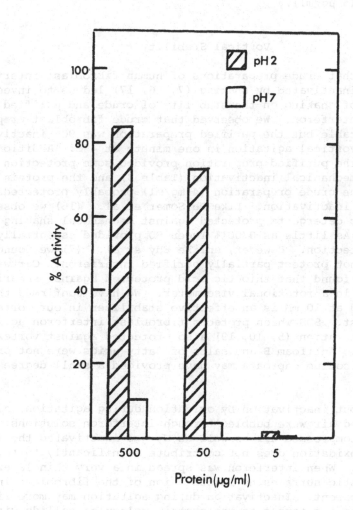

Figure 3

Effect of protein concentration upon stability of purified human
fibroblast interferon at 37°C. Interferons at the indicated protein
concentrations were stored at 37°C for 4 days in either 0.1 M sodium
phosphate (open) or 0.1 M KCl-HCl, pH 2 (crosshatched) and then
assayed for residual interferon activity.

pH 2 than at pH 7 for all protein concentrations tested, but even
at pH 2 the interferon became quite labile at very low protein concentrations (5 µg/ml).

Vortical Stability

Reports that crude preparations of human fibroblast interferons
are readily inactivated by shaking (7, 16, 17) led us to investigate
the effects of shaking on the stability of crude and purified
fibroblast interferon. We observed that crude fibroblast preparations were stable but the purified preparation was 90% inactivated
by vigorous vortical agitation in one minute at 4°C. Addition of
proteins to the purified preparation provided some protection
against the mechanical inactivation (Table 3) and the protein
present in the crude preparation (4 mg/ml) probably protected
against this inactivation. Like De Somer et al. (16) we observed
that nonionic detergents protected against mechanical shaking inactivation. As little as 0.001% Tween 80 provided essentially
complete protection. However, unlike Edy et al. (7) we found that
acid pH did not protect partially purified interferon. Cartwright
et al. (18) found that thioctic acid protected against shearing
inactivation in a rotational viscometer. We have confirmed that
thioctic acid at 10 mM is an effective stabilizer in our vortex
agitation test. SDS which protects fibroblast interferon against
thermal inactivation (5, 10, 19) also protected against vortex
inactivation. Antifoam B and salts of fatty acids were not protective, but sodium caproate may have provided a small degree of
protection.

To rule out inactivation by oxidation during agitation, nitrogen
and compressed air were bubbled through interferon solutions.
Bubbling of compressed air as well as N_2 gas inactivated the interferon; thus oxidation does not contribute significantly to this
inactivation. When interferon was spread in a very thin layer on
glass or plastic surfaces no inactivation of the fibroblast interferon was apparent. Inactivation during agitation may more likely
be due to shearing forces as the protein molecules collide with
the surface of the container during agitation. Cartwright et al.,
(8) found that the two factors controlling shear stress in the
rotational viscometer, rotational speed and the width of the annual
gap, had the predicted inactivating effects as if shearing were
the force responsible for the inactivation.

Freeze-dried Interferon

Freeze-dried preparations of crude fibroblast interferon are
much more stable than liquid preparations, retaining 50% of the
original activity even after 24 hours of heating at 90°C. Prepara-

Table 3

Stabilization Against Vortical Inactivation of Human Fibroblast
Interferon Purified by Immunoaffinity Chromatography

Additive	Percent residual activity after agitation
Control (0.1 M Na-phosphate, pH 7)	6
Control (0.1 M HCl-KCl, pH 2)	9
Cytochrome C, 100 µg/ml, pH 7	12
Cytochrome C, 500 µg/ml, pH 7	27
Ovalbumin, 500 µg/ml, pH 7	14
Human serum albumin, 500 µg/ml, pH 7	47
Bovine serum albumin, 500 µg/ml, pH 7	85
Tween-80, 0.5%, pH 7	100
Tween-80, 0.1%, pH 7	83
Tween-80, 0.05%, pH 7	84
Tween-80, 0.01%, pH 7	100
Tween-80, 0.001%, pH 7	85
Tween-80, 0.01%, pH 2	100
Tween-20, 0.1%, pH 7	52
Tween-20, 0.01%, pH 7	20
Antifoam B, 0.1%-0.0001%, pH 7	5
Dextran T-70, 5%, pH 7	12
Dimethyl adipimidate, pH 7	15
Glycerol, 1%-10%, pH 7	7
Na-Caproate, 0.1%, pH 7	22
Na-Oleate, 0.05%, pH 7	13
Sodium dodecyl sulfate, 0.01-0.1%, pH 7	100
Saline, pH 7	5
Thioctic acid, 1.0 mM, pH 7	2.8
Thioctic acid, 10 mM, pH 7	85

tions, retaining 50% of the original activity even after 24 hours of
heating at 90°C. Preparations at pH 2.0 and 7.0 had essentially the
same stability when freeze-dried.

Whereas added extraneous proteins did not further enhance the
stability of freeze-dried crude preparations, the addition of
extraneous proteins greatly endhanced the stability of freeze-dried
immunoaffinity purified interferon (Figure 4). The freeze-dried
interferon was heated in a linear nonisothermal test (14) in which
the temperature is gradually increased from 50°C to 90°C. The preparation with no added extraneous protein (containing 25 µg/ml cyto-

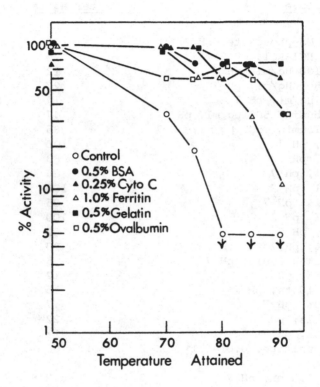

Figure 4

Linear nonisothermal accelerated storage test of freeze-dried
purified fibroblast interferon in pH 7 buffer with the indicated
proteins in addition to 25 µg of protein in the interferon preparation. Total volume before freeze-drying was 1.0 ml.

chrome C) had been nearly completely inactivated by the time the
temperature reached 80°C, whereas all of the preparations containing
added proteins still had antiviral activity after 90°C was attained.
Bovine albumin, cytochrome C, gelatin and ovalbumin provided essentially 100% stability whereas ferritin was less effective.

Predictive Test of Interferon Stability

There is a need for tests that reliably predict the shelf-life
of liquid interferon preparations. Multiple isothermal (MID) tests
utilizing at least three temperatures can be used to predict the
stabilities of suspensions of freeze-dried biological materials.
Briefly, this method is based on the observation that the logarithm
of potency of biological materials declines linearly at a selected
temperature in accordance with the Arrhenius theory. From Arrhenius
plots of inactivation rates at at least 3 elevated temperatures, the
inactivation rates during storage at low temperatures can be calculated. The data in Figure 5 shows that the MIS tests can predict
the stability at 37° of a freeze-dried murine interferon standard
that we have supplied to the National Institutes of Health. After
4 years of storage at 37° the standard still retained greater than
50% of its antiviral potency. The data in Table 4 indicate the stability of freeze-dried interferon standards supplied in the National
Institutes of Health predicted by long-term multiple isothermal
tests. When stored at 4°, both the mouse and human leukocyte standards are extremely stable, with predicted losses of 8% of activity
in 110 years for the mouse standard and 5% in 28.8 years for the
human leukocyte standard.

A more rapid test for screening candidate stabilizing materials
and predicting their stabilities, the linear nonisothermal (LNS)
accelerated storage test has been developed (14). This test involves
a programmed linear rate of increase of temperature with sampling
done as selected temperatures are achieved. This test also can be
related to the Arrhenius equation to make predictions of stabilities
of freeze-dried materials at any temperature.

High-temperature tests of liquid preparations do not seem
feasible for predicting stability in light of the anomolous results
observed at low and high temperature in the presence of chaotropic
salts. It seems that multiple isotermal tests at relatively low
temperatures (29° to 45°) may be the best way to predict the
shelf-life of liquid preparations, and long-term evaluations of
this test are now in progress.

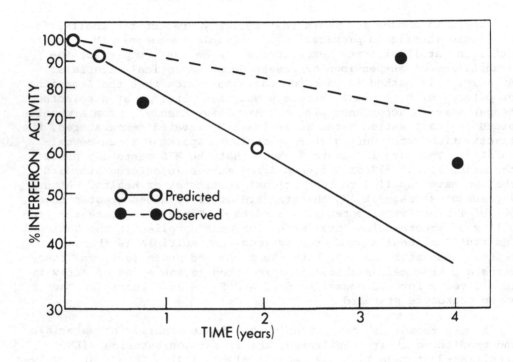

Figure 5

Confirmatory stability test of freeze-dried murine interferon
standard G002-904-511 at 37°. The predicted loss was determined
from multiple isothermal stability tests. The observed activity
after storage at 37° was expressed as the % of replicate samples
stored at -70° and titrated at the same time as those stored at
37°.

Table 4

Characteristics of Freeze-dried Interferon Standards
Supplied to the National Institutes of Health

NIH catalog number	Species	Virus Inducer	Inactiva-tion of inducer	Titer assigned	Predicted Stability (years to lose 1,000 units)	
					at 20°	at 4°
G002-904-511	Mouse	Newcastle disease	HClO$_4$	12,000	5.5	110
G023-901-527	Human (leukocyte)	Sendai	pH 2	20,000	1.4	28.8
G019-902-528	Rabbit	Bluetongue	pH 3.5	9,000	1.0	6.3

DISCUSSION

It is apparent that rapid inactivation of purified human fibroblast interferon results from heating and mechanical stresses. Agents which protect against thermal inactivation do not necessarily provide protection from mechanical stress inactivation, suggesting that there are indeed different forces working to inactivate the interferons. The inactivation due to heating may reflect changes in the covalent structure of proteins (20). Conditions which stabilize fibroblast interferons, such as low pH, instead of stabilizing the interferon in its native conformation may generate unfolded molecules which, upon subsequent cooling and neutralization or removal of the reagent, may refold to the biologically active form (15).

It appears that inactivation in the vortical agitation test is a result of shearing forces. Cartwright et al. (8, 18) suggest that shearing forces promote formation of molecular aggregates as a result of formation of intermolecular disulphide bonds. This hypothesis is compatible with the protective effects of Tween 80 or low pH (7, 16) since Tween 80 by its detergent action may prevent formation of intermolecular disulphide bonds and acid pH may prevent the random formation of such bonds. The effectiveness of thioctic acid at low concentrations and its low toxicity (18) make it one attractive stabilizer for use during the purification of human fibroblast interferon.

The use of freeze-dried preparations, of course, obviates the problem of shear inactivation. Accelerated storage tests such as the multiple isothermal test have established that freeze-dried preparations have suitably long shelf-lives for the long-term storage of interferon for clinical use.

However, in order to generate large quantities of purified stable interferon there is a need to establish an armamentarium of different nontoxic techniques and approaches that will enable investigators to stabilize fibroblast interferon at each step in the preparation, processing, concentration, purification, storage, shipment and clinical administration (21).

ACKNOWLEDGMENT

This work was supported by an award from the National Institutes of Health Nol AI 42520. We are grateful to Mary Dixon and Christine Schoenherr for their excellent technical assistance.

REFERENCES

1. Fantes, K.H. 1973. Purification and physico-chemical properties of interferons. Pgs. 171-200 in N.B. Finter, ed. Interferons and Interferon Inducers. North-Holland Publishing Co., New York.

2. Ng, M.H., and J. Vilcek. 1972. Interferons: physico-chemical properties and control of cellular synthesis. Advances in Protein Chemistry. 26:173-239.

3. Cesario, T.C., P.J. Schryer and J.G. Tilles. 1977. Relationship between the physiochemical nature of human interferon, the cell induced and the inducing agent. Antimicrobial Agents and Chemotherapy 11:291-298.

4. Mogensen, K.E. and K. Cantell. 1974. Human leukocyte interferon: a role for disulphide bonds. J. Gen. Virol. 22:95-103.

5. Stewart, W.E. II, P. De Somer, V.G. Edy, et al. 1975. Distinct molecular species of human interferons: requirements for stabilization and reactivation of human leukocyte and fibroblast interferons. J. Gen. Virol. 26:327-331.

6. Valle, M.J., G.W. Jordan, S. Haahr, et al. 1975. Characteristics of immune interferon produced by human lymphocyte cultures compared to other human interferons. J. Immunology 115:230-232.

7. Edy, V.G., A. Billiau, M. Joniau et al. 1974. Stabilization of mouse and human interferons by acid pH against inactivation due to shaking and guanidine hydrochloride. Proc. Soc. Exp. Biol. Med. 146:249-253.

8. Cartwright, T., O. Senussi and M.D. Grady. 1977. The mechanism of the inactivation of human fibroblast interferon by mechanical stress. J. Gen. Virol. in press.

9. Mogensen, K.E. and K. Cantell. 1973. Stability of human leukocyte interferon towards heat. Acta Path. Microbiol. Scand. Section B. 81:382-383.

10. Knight, E., Jr. 1976. Interferon: purification and initial characterization from human diploid cells. Proc. Nat. Acad. Sci. USA 73:520-523.

11. Berman, B. and J. Vilcek. 1974. Cellular binding characteristics of human interferon. Virology 57:378-386.

12.   Anfinsen, C.B., S. Bose, L. Corley, and D. Gurari-Rotman.
      1974.   Partial purification of human interferon by affinity
      chromatography.   Proc. Nat. Acad. Sci. USA 71:3139-3142.

13.   Jameson, P., M.A. Dixon, and S.E. Grossberg.   1977.   A sensi-
      tive interferon assay for many species of cells: encephalo-
      myocarditis virus hemagglutinin yield reduction.   Proc. Soc.
      Exp. Biol. Med. 55:173-178.

14.   Greiff, D. and C. Grieff.   1972.   Linear nonisothermal single-
      step, stability studies of dried preparations of influenza
      virus.   Cryobiology 9:34-37.

15.   Jariwalla, R.J., S.E. Grossberg and J.J. Sedmak.   1977.
      Effect of chaotropic salts and protein denaturants on the
      thermal stability of mouse fibroblast interferon.   J. Gen.
      Virol. 35:45-52.

16.   De Somer, P., M. Joniau, V.G. Edy et al.   1974.   Mass pro-
      duction of human interferon in diploid cells.   Pgs. 39-46 in
      C. Waymouth, ed.   The Production and Use of Interferon for
      the Treatment and Prevention of Human Virus Infections.
      The Tissue Culture Association, Rockville, MD.

17.   Havell, E.A. and J. Vilcek.   1974.   Mass production and some
      characteristics of human interferon from diploid cells.   Pg.
      47 in C. Waymouth, ed.   The Production and Use of Interferon
      for the Treatment and Prevention of Human Virus Infections.
      The Tissue Culture Association, Rockville, MD.

18.   Cartwright, T., O. Senussi and M.D. Grady.   1977.   Reagents
      which inhibit disulphide bond formation stabilise human
      fibroblast interferon.   J. Gen. Virol. in press.

19.   Vilcek, J., E.A. Havell, and S. Yamazaki.   1977.   Antigenic
      physico-chemical and biological characterization of human
      interferons.   Annals of the New York Academy of Sciences
      284:703-710.

20.   Tanford, C.   1978.   Protein denaturation.   Parts A and B.
      Adv. Prot. Chem. 23:122-275.

21.   Sedmak, J.J. and S.E. Grossberg.   1977.   Stabilization of
      Interferons.   Texas Reports on Biology and Medicine.   In press.

INTERFERON ASSAY ANOMALY VARIATION OF INTERFERON RESPONSE WITH

CELL TYPE AND SIALIC ACID CONTENT

A.A. Schwartz, Ph.D. and D. Villani-Price

Searle Laboratories

Skokie, Illinois U.S.A.

ABSTRACT

Human interferon derived from leukocytes and cultured fibroblast sources has been compared by assay in two human cell systems. The ratio of the activities of the two interferon preparations differed markedly in two assay systems. Neuraminidase treatment had no effect on the activity of either interferon in the more sensitive system but reduced the activity of the fibroblast material in the second system so that the resulting ratio of activities approximated that seen in the first assay system.

INTRODUCTION

Although variations in the sensitivity of human cells to human interferon are known to exist (1,2) and it has furthermore been observed that some non-human cells may be more sensitive to interferon than the human system (3), a human interferon standard (B 69/19) (4) is generally employed with the assumption that at least some relative activities can be measured. Recently, differences in the dose-response curves seen in different assay cells have been observed (5). We have compared the ratio of activity of preparations of human interferon from leukocyte and fibroblast sources on two different human cells and have found that the ratio of activities as conventionally determined is a function of the cells used to measure this activity. More particularly, we have shown that the residual activity of human fibroblast interferon after neuraminidase treatment is also dependent on the

153

cells in which this activity is measured.

Purified human leukocyte interferon (P-IF) and concentrated
human leukocyte interferon (C-IF), sendai virus induced, was
kindly provided by Dr. Kari Cantell, State Serum Institute,
Helsinki, Finland. Human fibroblast interferon (F-IF), poly I-
poly C induced, was supplied by Dr. T. Cartwright, Searle
Laboratories, High Wycombe, England. The FS-7 human fibroblast
cell strain and the vesicular stomatitis virus (VSV) were the
gifts of Dr. Jan Vilcek, New York University Medical School,
New York City; the U-amnion cell line was generously provided
by Dr. S. Baron, NIH, Bethesda. Interferon titres were measured
by the inhibition of cytopathic effect (CPE) following VSV
challenge (6). The FS-7 and the U cell assays differ only in
the virus dose. The FS-7 cells are infected at .07 pfu per cell;
the U cells at 10 pfu per cell and the latter read at 24 rather
then 48 hours post challenge. We have found that this procedure
gives the same results as the lower dose/longer time method
in these cells. Neuraminidase was purchased from Schwarz/Mann,
lot #8Z2199. The enzyme was free of protease activity in an
assay sensitive to 1.5 international units of standard protease/
ml (7). Samples for neuraminidase treatment were dialyzed
against 0.15 M NaCl, 0.05M sodium acetate-acetic acid buffer,
20 mM ca Cl_2 pH 5.5 (8). Reactions were carried out at 37°,
4 hours, in the presence of 5 µl chloroform with a total of 35
international units (IU) neuraminidase per mg protein added in
two parts, at 0 and 2 hours. Incubated controls were treated
identically except for the enzyme addition.

The results of assaying the interferons from two different
sources on each of the two cell lines are presented in Table I.
The P-IF standard is an internal standard made by diluting P-IF
with 10 mg/ml bovine serum albumin (BSA) in phosphate buffered
saline (PBS) pH 7.2, aliquoting and freezing at -70° until use.
The difference in sensitivity of the two cell lines to interferon
was not unexpected. It is, however, also clear that the ratio
of activities of the two preparations differ markedly when
assayed in the two different cell lines. The titre of fibroblast
interferon is 10 fold lower measured against a leukocyte standard
when the assay is done in FS-7 cells despite the fact that this
more sensitive system yields apparent titres that are as much as
1000 x higher. The variations in absolute titre in the FS-7
system are attributed to a characteristic increase in sensitivity
of these cells as a function of higher passage number (J. Vilcek,
personal communication)(9).

In Table II, we present the results of studies on the effect
of neuraminidase on this phenomenon. Samples incubated in the
absence of neuraminidase are used as a control. It will be

TABLE I

Titration of leukocyte and fibroblast interferon on 2 human cell systems

	Sample	Titre	Sample	Titre	Ratio P-IF/F-IF
U-amnion cells	P-IF std	25,000	F-IF	360	69
"		84,000	"	1,200	70
"		52,000	"	860	60
FS-7 cells	P-IF std	3.3×10^6	F-IF	2,900	1138
"		8.2×10^6	"	8,192	1000
"		33×10^6	"	28,000	1178

Paired assays on one line done at the same time on the same cell preparation.

TABLE II

Titration of neuraminidase treated and control interferons in 2 human cell systems.

Interferon	Assay Cell	Dialyzed Control	Incubated Control	Neuraminidase treated
F-IF	FS-7	16,000	16,000	12,600
F-IF	FS-7	12,000	12,000	8,000
P-IF	FS-7	9.6×10^6	10.4×10^6	9.6×10^6
F-IF	U-amnion	1,880	940	60
F-IF+	U-amnion	1,200	760	67**
P-IF std+	U-amnion	84,000	71,000	84,000**
F-IF*	U-amnion	430	432	87
P-IF*	U-amnion	1.3×10^6	1.3×10^6	1.3×10^6
C-IF*	U-amnion	9,230	6,529	9,230

+,* Simultaneous assay on the same cell preparation

** P-IF std (neur)/F-IF (neur) = 1235

noted that neither the leukocyte or fibroblast material show any
evidence of loss of activity when assayed on FS-7 cells. The
same is true of leukocyte material assayed on U cells. The
fibroblast interferon shows a marked loss of activity under the
latter conditions. The ratio between the measured activity
of the leukocyte and fibroblast material is however not characteristic of that seen when assayed on U cells but is comparable
to that seen when assayed on the FS-7 cells.

Similar results are obtained with P-IF (without BSA diluent)
and C-IF indicating that neither the diluent nor the purification
affect the comparison.

The differing ratios derived in Figure 1 indicate that while
it may be appropriate to use a single leukocyte preparation as a
means of standardizing the assay of leukocyte interferon in
different laboratories, this will not be a valid procedure for
fibroblast material unless the same cells are used. Furthermore,
it appears that in some as yet unexplained way, the sialic acid
component of fibroblast interferon may play a major role in its
action on U-amnion cells.

One of the enduring problems of interferon research has been
the difficulty of standardization. We believe that these data
point to one source of such difficulty. The apparent ability
of neuraminidase treatment to eliminate a biological distinction
between leukocyte and fibroblast interferon (the heightened
activity of the latter vis a vis the former on U-amnion cells)
suggests that the carbohydrate moieties are an appropriate place
to look for the basis of this anomaly. This view is supported
by a recent report (10) suggesting that leukocyte interferon has
little or no carbohydrate content while fibroblast-derived
material is clearly a glycoprotein. In addition, we have
observed that the isoelectric focusing pattern of fibroblast
interferon shifts markedly after neuraminidase treatment while
the change in the leukocyte material is present but much smaller
(unpublished). (E. Havell, personal communication) (11).

We wish to thank Dr. S. Papaioannou and Ms. M. Babler who
ran the protease assay and our colleagues mentioned earlier who
provided the interferon and cell lines used in this work.
A. A. Schwartz, D. Villani-Price, Searle Laboratories, Skokie, Ill.
U.S.A.

REFERENCES

1. Finter, N.B.   1966.   Interferon assays and standards.
    Pgs. 87-118 in N.B. Finter, ed.   Interferons.   North Holland
    Publishing Co., Amsterdam.

2. Dahl, H.   1973.   Micro assay for mouse and human interferon
    .2.   Dose response in different cell-virus systems.   Acta
    Path. Microbiol. Scand. Section B 81:359-364.

3. Gresser, I., M.T. Bundu, D. Brouty-Boye, and M. Tovey.   1974.
    Pronounced antiviral activity of human interferon on bovine and
    porcine cells.   Nature 251:543-545.

4. Symp. Ser. Immunobiol. Stand 14:326-328 (Karger, Basel 1970).

5. Edy, V.G., A. Billiav., and P. DeSomer.   1976.   Human
    fibroblast and leukocyte interferons show different dose-
    response curves in assay of cell protection.   J. Gen. Virol.
    31(2):251-255.

6. Havell, E.A. and J. Vilcek.   1972.   Production of high-
    titered interferon in cultures of human diploid cells.
    Antimicrob. Ag. Chemother. 2:476-484.

7. Alkjaersig, N., A.P. Fletcher and S. Sherry.   1959.
    Epsilon - Aminocaproic acid: an inhibitor of plasminogen
    activation.   J. Biol. Chem. 234:832-837.

8. Viti, A., V. Bocci, M. Russi and G. Rita.   1970.   Effect of
    neuraminidase on rabbit urinary interferon.   Experentia 26(4):
    363-364.

9. J. Vilcek, personal communication.

10. Jankowski, W.J., M.W. Davey, J.A. O'Malley E. Sulkowski and
    W.A. Carter.   1975.   Molecular structure of human fibroblast
    and leukocyte interferons: Probe by lectin and hydrophobic
    chromatography.   J. Virol 16(5):1124-1130.

11. E. Havell, personal communication.

INTERFERON THERAPY FOR NEOPLASTIC DISEASES IN MAN IN VITRO AND
IN VIVO STUDIES

Stefan Einhorn and Hans Strander

Radiumhemmet, Karolinska Hospital
S-104 01 Stockholm 60, Sweden

SUMMARY

With the object of examining the anti-tumour effect of
exogenous interferon therapy in man a research programme has been
initiated at the Karolinska Hospital.

Established cell lines obtained from patients with Burkitt's
and other types of lymphoma, leukaemia, osteosarcoma, mammary
carcinoma and fibrosarcoma and from fibroblast cultures displayed
a variable sensitivity to the cell multiplication inhibitory
activity of interferon. All the monolayer cultures tested were
found to be sensitive to interferon at concentrations between
10 and 300 units/ml. Some lymphoma cell lines were not sensitive
to interferon even at concentrations as high as 10.000 units/ml,
while others were sensitive at concentrations between 2 and 300
units/ml. The interferons tested appeared to show a degree of
tissue specificity.

Controlled studies in vivo are being performed on osteosarcoma,
juvenile papilloma of the larynx, multiple myeloma and small-cell
carcinoma of the lung. The clinical results of this research
obtained to date, together with the results obtained in model
experiments, would appear to warrant accelerated production of
human interferon.

INTRODUCTION

At the Karolinska Hospital interferon therapy was first
introduced in 1969. Since then a fund of experience of administra-

tion, metabolism, side effects and therapeutic results has been
analysed and reported (for references see 4, 18, 19). The human
leukocyte interferon used in all these studies was prepared by
Dr. Kari Cantell at the Central Public Health Laboratory,
Helsinki, Finland. The methods of production of purification of
this interferon have been described in detail (for survey and
references see 15).

Of interferons, leukocyte, fibroblast and immune types are
the 3 main ones that have been described to date (12, 14, 26).
In its antigenic properties interferon produced by lymphoblastoid
cells resembles leukocyte interferon (24) but it also seems to
consist partly of fibroblastlike interferon (25). It has been
demonstrated in animals that interferon is capable of inhibiting
the growth of various kinds of tumours, but the underlying
mechanisms remain obscure. Various interferon preparations have
been used in these studies (for references see 8, 9).

The object of the investigations with which this article is
concerned was to examine in vitro one particular property of
interferon, namely, its capacity to inhibit cell multiplication
(11, 16). Results obtained to date on established human cell
lines raise the question whether the anti-tumour effects exerted
by different interferons might depend on the origin of the tumour
cells tested. The findings obtained so far in vivo are also
outlined.

MATERIAL AND METHODS

Cell Lines

The cell lines were obtained from Professor Georg Klein
(Department of Tumor Biology, Karolinska Institute, Stockholm,
Sweden), Professor Jan Pontén (The Wallenberg Laboratory, Uppsala,
Sweden), Dr. Kenneth Nilsson (The Wallenberg Laboratory, Uppsala,
Sweden), Dr. Jørgen Fogh (Sloan-Kettering Institute for Cancer
Research, New York, USA), and Dr. Beppino Giovanella (The Stehlin
Foundation for Cancer Research, Houston, Texas, USA).

Interferon Preparations

Human leukocyte interferon was obtained from Dr. Kari
Cantell (Central Public Health Laboratory, Helsinki, Finland).
Its mode of preparation has been described elsewhere (5); its
specific activity was about 3×10^6 units/mg of protein. Human
fibroblast interferon was prepared by Dr. Ed Havell and Dr.
Jan Vilcek (New York University Medical Center, New York, USA),

(13). It was kindly provided by them; its specific activity was
1.5×10^4 units/mg of protein.

Interferon Assay

The interferons were assayed in U cells by vesicular stomatitis
virus (VSV) plaque reduction as described earlier (20). The
activity of the interferon is expressed in International Units
(human standard reference 69/19).

Cell Growth Experiments

The cell lines were grown either as suspension cultures
(lymphoid cell lines) or as monolayer cultures; the techniques
have been described earlier (7, 23). The cultures were split
once a week so as to contain either 2×10^5 (lymphoid cell lines)
or 5×10^4 (monolayer cultures) living cells per ml.

Definition of Interferon Sensitivity

The sensitivity of a cell line to interferon is defined
as the concentration of interferon required to cause a 50 per
cent inhibition of cell multiplication in 2 weeks.

Clinical Trials

Details of the clinical trials have been given in a series of
articles (for references see 18, 19).

RESULTS

In Vitro Studies on the Cell Multiplication Inhibitory
(CMI) Activity of Interferon

Burkitt's lymphoma cell lines. The Burkitt's lymphoma cell
lines could be assigned to 2 groups according to their sensitivity
to the CMI activity of the leukocyte interferon: one group
showed a 50 per cent inhibition of growth in 2 weeks at an
interferon concentration of between 2 and 300 units/ml; the other
group did not exhibit this level of inhibition at a concentration
of 10.000 interferon units/ml (Table]). These results confirms
and supplements earlier findings at this Laboratory (1).

Other lymphoid cell lines. A similar division into 2 groups

TABLE 1

Sensitivity of Burkitt's lymphoma cell line to inhibition of
cell multiplication by human leukocyte interferon.

Cell Line	Sensitive to [a] (units/ml)
DAUDI	2
P$_3$HR-1	10
NALIAKA	30
ODOUR	200
SERAPHINE	300
PESS	10,000
AKUBA	10,000
RAJI	10,000
LY-46	10,000
NAMALWA	10,000
MAKU	10,000

[a] Expressed as the interferon concentration required to produce
a 50 per cent inhibition of growth in 2 weeks.

could be made for lymphoid cell lines other than those of
Burkitt's lymphoma origin the respective ranges of interferon
concentration then being 10-300 units/ml and above 10.000 units/ml
(Table 2).

    Osteosarcoma cell lines. No assignment to 2 distinct
sensitivity groups could be made for the osteosarcoma cell lines,
all those tested being sensitive to leukocyte interferon at
concentrations between 10 and 300 units/ml (Table 3).

    The observed difference in the degree of sensitivity of the

TABLE 2

Sensitivity of various non-Burkitt lymphoid cell lines to the
inhibition of cell multiplication by human leukocyte interferon.

Cell line	Origin of cell line	Sensitive to [a] (units/ml)
266BL	Myeloma	10
MOLT	Leukaemia	10
715	Lymphocytic lymphoma	65
IHTC-33	Infectious mononucleosis	100
RPMI-7101	Infectious mononucleosis	100
61M	Lymphoblastoid from normal subject	100
718	Lymphoblastoid from normal subject	100
RPMI-7305	Lymphoblastoid from normal subject	100
SKL-1	Leukaemia	300
698M	Lymphocytic lymphoma	>10,000
DSTC-4	Infectious mononucleosis	>10,000
(HE)-6410	Leukaemia	>10,000
NC-37	Lymphoblastoid from normal subject	>10,000

[a] See footnote to Table 1.

TABLE 3

Sensitivity of osteosarcoma cell lines to the inhibition of
cell multiplication by human leukocyte interferon.

	Sensitive to [a] (units/ml)
Cell line	
SAOS-2	10
SI-II	20
$T_2 56$	30
TE-85	30
393T	100
2T	100
20S	100
SI-I	300
$T_2 278$	300

[a]See footnote to Table 1.

various osteosarcoma cell lines might at least partly be explained
by the fact that the cell line 2T displayed a different degree of
sensitivity to interferon at two passage levels tested (56th
and 278th) (23).

Other monolayer cultures. For these lines no division into
2 groups could be made. Three mammary carcinoma cell lines and
one fibrosarcoma cell line were all sensitive to leukocyte interferon at concentrations of 10 - 300 units/ml (Table 4). Three
fibroblast cell lines were classed as sensitive, concentrations of
between 100 and 300 units/ml being required to cause a 50 per cent
inhibition of growth in 2 weeks. This result supplements earlier
findings made at this Laboratory (21).

Attempts to select an interferon-resistant cell line. Seven
osteosarcoma cell lines grown in the presence of interferon for
8 weeks displayed no significant change in sensitivity to interferon

TABLE 4

Sensitivity of various monolayer cultures to the inhibition of
cell multiplication by human leukocyte interferon.

	Origin	Sensitive to [a] (units/ml)
Cell line		
BT-20	Mammary carcinoma	10
MCF-7	Mammary carcinoma	100
MDA-MB-231	Mammary carcinoma	100
SW-684	Fibrosarcoma	100
GM-258	Fibroblast	100
393S	Fibroblast	100
S2	Fibroblast	300

[a]See footnote to Table 1.

over this period (Figure 1). One cell line (SI-II), grown in the
presence of interferon for 6 months, likewise displayed no change
in sensitivity with time. Seven of the lymphoid cell lines that
had proved to be sensitive to interferon were cultured in the
presence of interferon for 8 weeks without exhibiting a significant
change in sensitivity.

Is interferon tissue-specific? The CMI activities of human
leukocyte and fibroblast interferons were compared on osteosarcoma
and Burkitt's lymphoma cell lines (7). The osteosarcoma cell
lines tested (SAOS-2 and SI-II) were more sensitive to fibroblast
than to leukocyte interferon. The Burkitt's lymphoma cell lines
(Daudi and P3HR-1), on the other hand, were more sensitive to
leukocyte than to fibroblast interferon (Table 5). These studies
are at present being extended to include other cell lines.

In Vivo Studies on the Efficacy of Interferon Therapy

Osteosarcoma. In a clinical trial begun at the Karolinska
Hospital in 1971 osteosarcoma patients are being given interferon
as the sole form of adjuvant therapy (22). To date the series

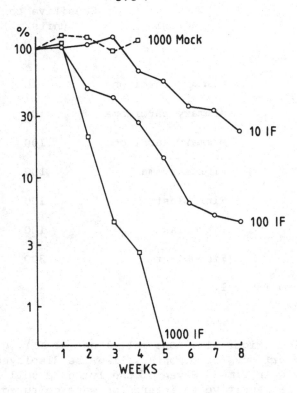

Figure 1

Cell multiplication inhibitory effect of longterm interferon
treatment at doses from 10 to 1000 interferon units/ml on the
human osteosarcoma cell line 393T. The mock preparation was
produced as the interferon but without the addition of an inducing
virus. One thousand "mock" units contain the same quantity
of protein as 1.000 interferon units.

TABLE 5

Percentage inhibition of cell multiplication in osteosarcoma and Burkitt's lymphoma cell lines after 2 weeks' exposure to interferon at the concentrations 10 and 100 units/ml. The percentages were calculated from the expression

$$100 - \frac{\text{counts of living cells in interferon dishes}}{\text{control}} \times 100$$

| | Osteosarcoma | | Burkitt's lymphoma | | Cell lines compared |
	Leukocyte interferon	Fibroblast interferon	Leukocyte interferon	Fibroblast interferon	
10 interferon units/ml	27	51	82	19	SI-II vs.Daudi
	38	94	6	0	SAOS-2 vs.P_3HR-1
100 interferon units/ml	87	100	100	100	SI-II vs.Daudi
	83	100	69	28	SAOS-2 vs.P_3HR-1

comprises 28 consecutive patients. The treatment consists in one
intramuscular injection of 3 x 10^6 units of leukocyte interferon
daily for one month, during which time the patient remains in
hospital and is operated upon (exarticulation, amputation or
resection]. Therapy is then continued on an ambulatory basis with
3 interferon injections a week for the next 17 months. Two control
groups have been collected, one consisting of 35 historical
patients receiving treatment at the Karolinska Hospital between
1952 and 1971, and the other group comprising all 23 osteosarcoma
patients treated at other Swedish hospitals during the period
1972 - 1974. Neither control group has received adjuvant therapy.

To ascertain whether the tumours in the experimental and
the two control groups were equivalant, all the patients (except
for the last 7 given interferon) were reviewed in detail by a team
of pathologists and clinicians with experience of bone tumours
(2). This review showed that the prognosis is less favourable
for the historical group than for the other 2 groups. At present,
therefore, the group receiving interferon therapy can be compared
with the concurrent control group only.

Life table analysis (17) revealed that 64 per cent of the
patients given interferon were free from metastases after
2 1/2 years, compared with 30 per cent in the concurrent control
group (Strander, Aparisi, Brostrom, Cantell, Ingimarsson,
Jacobsson, Lagergren, Nilsonne & Soderberg, unpublished observations). Life table analysis of survival showed that after 2 1/2
years 73 per cent of the interferon group were still living,
against 35 per cent in the concurrent control group.

Hodgkin's disease. In one patient with Hodgkin's disease
interferon therapy was effective at first, but subsequently
growth of the tumour was resumed (3).

Multiple myeloma. Two patients with multiple myeloma were
given leukocyte interferon as the sole form of treatment. The
first patient, a woman with a melphalan-resistant IgG-producing
myeloma, received 6 x 10^6 units daily. A beneficial result of
the therapy was reflected in a clinical improvement, favourable
effects on laboratory parameters, a fall in the serum level of
tumour-specific IgG, and a decrease in the number of tumour cells
in the bone marrow. Again, progression was revealed after initial
regression (Idestrom, Killander, Cantell and Strander, unpublished
observations). The second patient is under treatment with 3 x 10^6
units daily (Mellstedt & Strander, unpublished observations).

Juvenile laryngeal papillima. Two patients aged 5 and 12
years with juvenile laryngeal papilloma were given interferon
therapy at the Karolinsa Hospital (Haglund, Lundkvist, Strander,

Cantell & Wersäll, underline(unpublished observations). Both had previously
undergone repeated surgery. Prior to the interferon treatment the
introitus of the larynx was occluded by papilloma. Interferon was
given to one patient for 12 months and to the other for 6 months;
the dose was 3×10^6 units intramuscularly 3 times a week. Both
these patients responded favourably to this treatment showing
regression of their tumors after one month of interferon treatment.
Additional patients are being included in this study.

DISCUSSION

The purpose of the investigations reported her is to develop
an efficient form of treatment for neoplastic disease in man using
currently available interferon preparations. In the laboratory we
have so far focused our attention on the capacity of various interferon preparations to inhibit cell multiplication. Whether the CMI
activity has any therapeutic value has yet to be ascertained.
The discovery that the Burkitt's lymphoma cell lines could be divided
into 2 groups according to their sensitivity to interferon has
interesting implications. If interferon therapy proves effective
in this disease sensitivity tests might be performed in order to
select patients with sensitive tumours for therapy since sensitivity
tests can be performed on biopsy cells (Ernberg, Einhorn and
Strander, unpublished observations).

The osteosarcoma and mammary carcinoma cell lines were
sensitive to interferon at concentrations that can be achieved in
the serum of patients given interferon treatment (Table 6). These
diseases would be suitable for interferon therapy if CMI effects
are clinically important.

In 2 patients, one with Hodgkin's disease and the other with
multiple myeloma, there was progression of the tumours after an
initial period of regression during interferon therapy; whether
this renewed growth was due to the development of tumour cell
resistnace to interferon is not known. So far we have been unable
to detect interferon-resistant cells in any of the established
sensitive cell lines tested at this Laboratory. In experiments
employing animal cell populations this has however been possible
(10).

Experiments are at present being conducted with a view to
establishing whether antibodies to interferon are produced during
interferon therapy. The results of preliminary studies on the
effect of interferon on lymphocytes isolated from patients
receiving interferon indicate that these cells are as sensitive
to interferon at the end of longterm interferon treatment as they
were at the beginning (unpublished observations), but whether

TABLE 6

Tumour diseases studied at the Karolinska Hospital for evaluation
of efficacy of interferon therapy.

	Number of patients treated
Osteosarcoma	28
Multiple myeloma	2
Juvenile papilloma of the larynx	2
Hodgkin's disease	1
Total	33

this also holds true for the tumour cells in the same patients
is not known.

The CMI effect seems to display some measure of specificity
in respect of interferon preparation and the established cell line
tested, but whether this has any implications in vivo remains to
be explored in both animals and man.

Should we continue to use large amounts of interferon in
clinical trials on patients with neoplastic diseases? To produce
the quantities required is difficult and expensive. Should not
priority be given to antiviral forms of treatment therapeutic
effects of which have been reported (see this Workshop)? The results obtained to date at the clinical level would appear to
justify continuing clinical trials on human tumour disease. It
is evident then that extended large scale production of interferon
is warranted. At the Karolinska Hospital investigations are being
conducted on osteosarcoma, juvenile papilloma of the larynx,
multiple myeloma and small-cell carcinoma of the lung. Controlled
and randomized trials are being initiated. It will be a long
time before these studies yield definitive results; meanwhile
the expected enlarged production and further purification of
interferon will provide a more suitable product for clinical
application.

ACKNOWLEDGEMENTS

It is a great pleasure for us to acknowledge our indebtedness

to Professor Kari Cantell for generous supplies of human leukocyte
interferon, for assaying antiviral activities of the various
interferons and for many stimulating discussions. The clinical
studies summarized here would have been impossible without the
generous support of the Finnish Red Cross Blood Transfusion
Service.

We are also indebted to all the clinicians, nurses and
laboratory technicians who have participated in these studies.

Financial support for the investigations reported here was
made available by the Cancer Society of Stockholm, The Swedish
Cancer Association, The Karolinska Institute and the Karolinska
Hospital (Dagmar Hasselgren's Fund).

REFERENCES

1.   Adams, A., H. Strander, and K. Cantell. 1975. Sensitivity
     of the Epstein-Barr virus transformed human lymphoid cell
     lines to interferon. J.gen.Virol. 28:207-217.

2.   Aparisi, T., L.A. Broström, S. Ingimarsson, C. Lagergren,
     U. Nilsonne, H. Strander, and G. Söderberg. 1977. Prog-
     nostic factors in osteosarcoma. Can historical controls
     be used in current clinical trials? Intern.J.Rad.Oncol.Biol.
     Phys., in press.

3.   Blomgren, H., K. Cantell, B. Johansson, C. Lagergren, U.
     Ringborg, and H. Strander. 1976. Interferon therapy in
     Hodgkin's desease: Acta med.Scand. 199:527-532.

4.   Cantell, K. 1977. Prospects for the clinical use of exogenous
     interferon. Med.Biol., in press.

5.   Cantell, K., S. Hirvonen, K.E. Mogensen, and L. Pyhälä. 1974.
     Human leukocyte interferon: production, purification, stabilit
     and animal experiments. In: C. Waymouth (Ed.), The Production
     and Use of Interferon for the Treatment and Prevention of
     Human Virus Infections, Proceedings of a Tissue Culture
     Association Workshop. Tissue Culture Association, Rockville,
     pp. 35-38.

6.   Cantell, K., L. Pyhälä, and H. Strander. 1974. Circulating
     human interferon after intramuscular injection into animals
     and man. J.gen.Virol. 22:453-455.

7.   Einhorn, S., and H. Strander. 1977. Is interferon tissue
     specific? - Effect of human leukocyte and fibroblast
     interferons on the growth of lymphoblastoid and osteosarcoma
     cell lines. J.gen.Virol. 35:573-577.

8.   Gresser, I. 1972. Antitumour effects of interferon. Adv.
     Cancer Res. 16:97-140.

9.   Gresser, I. 1977. Antitumour effects of interferon. In:
     F. Becker (ed.), Cancer. A comprehensive treatise, in press.

10. Gresser, I., M.T. Bandu, and D.Brouty-Boyé. 1974. Interferon
     and cell division. IX. Interferon-resistant L1210 cells:
     characteristics and origin. J.Nat.Cancer Inst. 52:553-559.

11. Gresser, I., D. Brouty-Boyé, M. T. Thomas, and A. Macieira-
     Coelho. 1970. Interferon and cell division II. Influence of
     various experimental conditions on the inhibition of L1210
     cell multiplication in vitro by interferon preparations.

J.Nat.Cancer Inst. 45:1145-1153.

12. Havell, E.A., B. Berman, C.A. Ogburn, K. Berg, K. Paucker, and J. Vilcek. 1975. Two antigenically distinct species of human interferon. Proc.Nat.Ac.Sci.US 72:2185-2187.

13. Havell, E.A., and J. Vilcek. 1972. Production of hightitred interferon in cultures of human diploid cells. Antimicrob. Ag. Chemother. 2:476-484.

14. Jankowski, W.J., M.W. Davey, J.A. O'Malley, E. Sulkowski, and W.A. Carter. 1975. Molecular sturcture of human fibroblast and leukocyte interferons:probe by lectin and hydrophobic chromatography. J.Virol. 16:1124-1130.

15. Mogensen, K.E. and K. Cantell. 1977. Production and preparation of human leukocyte interferon. Pharmacology and therapeutics, in press.

16. Paucker, K., K. Cantell, and W. Henle. 1962. Quantitative studies on viral interference in suspended L cells. III. Effect of interfering viruses and interferon on the growth rate of cell. Virology 17:324-334.

17. Peto, R., M.C. Pike, P. Armitage, N.E. Breslow, D.R. Cox, S.V. Howard, N. Mantel, K. McPherson, J. Peto, and P.G. Smith. 1977. Design and analysis of randomized clinical trials requiring prolonged observation of each patient. II. Analysis and examples. Br.J.Cancer 35:1-39.

18. Strander, H. 1977. Anti-tumour effects of interferon and its possible use as an anti-neoplastic agent in man. Texas Reports Biol.Med., in press.

19. Strander, H. 1977. Interferons: Anti-neoplastic drugs? Blut, in press.

20. Strander, H., and K. Cantell. 1966. Production of interferon by human leukocytes in vitro. Ann.Med.Exp.Biol.Fenn. 44:265-273.

21. Strander, H., and K. Cantell. 1974. Studies on antiviral and antitumour effects of human leukcyte interferon in vitro and in vivo. In: The C. Waymouth (Ed.), The Production and Use of Interferon for the Treatment and Prevention of Human Virus Infections, Proceedings of a Tissue Culture Association Workshop. Tissue Culture Association, Rockville, pp. 49-56.

22.   Strander, H., K. Cantell, S. Ingimarsson, P.A. Jakobsson,
      U. Nilsonne, and G. Söderberg. 1977. Interferon treatment
      of osteogenic sarcoma - a clinical trial. In: M.A. Chirigos
      (Ed.), Modulation of host immune resistance in the prevention
      or treatment of induced neoplasias, Dec. 9-11, 1974. Fogarty
      Int. Center Proc., US Government Printing Office, Washington
      DC, No. 28, pp. 377-381.

23.   Strander, H., and S. Einhorn. 1977. Effect of human leukocyte
      interferon on the growth of human osteosarcoma cells in tissue
      culture. Int. J. Cancer 19:468-473.

24.   Strander, H., K.E. Mogensen, and K. Cantell. 1975. Production
      of human lymphoblastoid interferon. J. clin. Microbiol.
      1:116-117.

25.   Vilcek, J. 1977. Protein chemistry and molecular biology of
      interferons: a secular view. In: M.Revel, D. Gurari-Rotman,
      and E. Winoccour (Eds.), Symposium of interferons and the
      control of cell-virus interactions, Rehovot, Israel, May 2-6.

26.   Youngner, J.S., and S.B. Salvin. 1973. Production and
      properties of migration inhibitory factor and interferon in
      the circulation of mice with delayed type hypersensitivity.
      J. Immunol. 111:1914-1922.

RESULTS OF A FIVE-YEAR STUDY OF THE CURATIVE EFFECT OF DOUBLE STRANDED RIBONUCLEIC ACID IN VIRAL DERMATOSES AND EYE DISEASES

L. Borecky,[1] J. Buchvald,[2] E. Adlerova,[3] I. Stodola,[4]
E. Obrucnikova,[5] Z. Gruntova,[6] V. Lackovic,[1] and J.
Doskocil[7]

[1]Institute of Virology, Slovak Academy of Sciences,
  Bratislava;
[2]Dermatological Clinic, University Hospital, Bratislava
[3]Children's Clinic, Postgraduate School of Medicine,
  Bratislava
[4]County Hospital, Banska Bystrica
[5]Eye Clinic, University Hospital, Olomouc
[6]School of Pharmacy, Comenius University, Bratislava
[7]Institute of Organic Chemistry and Biochemistry,
  Czechoslovak Academy of Sciences, Prague

ABSTRACT

Double stranded RNA obtained from non-permissive E. coli cells
infected with f2-phage was tested in 5 hospitals in viral dermatoses such as herpes simplex recidivans, herpes zoster, male geni-
keratonconjunctivitis herpetica and conjunctivitis lignosa. The
results of clinical tests indicate that the preparation of phage
double stranded RNA applied topically is harmless for man and has,
in the majority of cases, a beneficial effect in the disease. This
conclusion was based on the judgment of physicians, opinion of
patients expressed in a questionnaire, and, results of a double-
blind experiment.

INTRODUCTION

In addition to the interferon/IF/inducing capability, at least
three activities are associated with the administration of double
stranded ribonucleic acid/ds RNA/to the animal organism. First,
both low and high molecular weight preparations exert an amplifying
effect on antibody production and cell-mediated antibody response
(1, 2, 3). Second, ds RNA's and their degradation products have a

175

regulatory effect on cell performance through antimitotic, metabolic and other effects (4,5). Third, ds RNAs may exert a probably non-IF mediated antiviral and anticancerous effect (6,7,8).

Since native ds RNAs are comparable in these activities with synthetic preparations and, moreover, all so far tested native ds RNAs proved to be IF-inducers while synthetic preparations differ in this respect, clinical testing of native ds RNAs finds additional justification (9,10,11,12).

The f2-phage double stranded RNA

An economically advantageous method for obtaining native ds RNA was developed in Czechoslovakia by Doskočil et al., (11). The method utilizes the pathological accumulation of replicative ds RNA in non-permissive F^+ E. Coli cells infected with a suppressor sensitive (amber)mutant of phage f2. As a result of incomplete multiplication cycle, in a few hours after infection, the bacterial cells become filled with ds RNA (f2-RNA) which may represent up to 40% of the cell content. For extraction and purification of ds RNA from infected cells, the method of Billeter and Weissman (13) and, alternatively of Franklin (14) were used. An additional treatment of the extracted phage ds RNA with 2-methoxyethanol according to Kirby (15) proved useful for removal of lipopolysaccharides. The preparation was characterized by sedimentation coefficient, thermal denaturation profile, and, polyacrylamide gel electrophoresis (16). In special cases, the derivative pulse-polarography was used for detection of minute amounts of single stranded nucleotides in preparation (17). The sedimentation coefficient of the major components of batches prepared in years between 1971 and 1973 was between 6S and 9S, and over 10S in those prepared after May 1973. On average, 100 mg of f2-RNA can be obtained from 1 liter of cultivation medium, and, the production costs of this ds RNA are significantly lower than in the case of synthetic polynucleotides. The yield and the physical parameters of f2-RNA are influenced, however, by the biological properties of the producer cells as well as the f2-phage used for infection.

As reported earlier, f2-RNA induces IF in tissue cultures and mice, and, protects them against encephalomyocarditis and tick-borne encephalitis (Hypr-strain) virus (18). In addition, it depresses the multiplication of tick-borne encephalitis and Sindbis virus in immunosuppressed mice (19,20).

Pharmacological Tests With f2-RNA

A crucial problem that hinders the therapeutical evaluation of ds RNAs is their potential toxicity for man (21). This consideration led us to limit our plans of evaluation of f2-RNA to diseases which require topical treatment. The pharmacological tests required for admission of f2-RNA to clinical trials in dermatological and eye diseases are summarized on Table I. In contradistinction to synthetic preparations, the purified f2-RNA showed no enhanced toxicity for mice sensitized agains poly I:C by the procedure developed by De Clercq et al., (22), and, proved to be less pyrogenic also for rabbits (23). Furthermore, the purified f2-RNA with a medium Sw 7.25 after prolonged treatment exerted a beneficial effect on NZB/Swiss mice hybrids which at age of 4-6 months develop regularly a SLE-like autoimmune disease. In consequence, the hydrids had a prolonged survival time and their migration indices in the presence of DNA- and RNA- antigens show normalization (24).

Table I

Pharmacological Tests with Phage Double-Stranded RNA

Tests with "Crude" and "Purified" Preparations of f2-RNA

1. Acute Toxicity/LD_{50} in mg/kf/   MOUSE   RAT

		MOUSE	RAT
Route:	intravenous	>50/36/[x]	>100
	intraperitoneal	>75/48/	≈120
	subcutaneous	≈250	>400

    Toxicity for NDV-Sensitized Mice[xx]: >50 mg/kg

2. Beneficial Effect in NZB/Swiss Mice
3. Pyrogenicity for Rabbits 10/0.1/[x] mg/kg
4. No Mutagenicity for **E. coli** : /in range 0.0002-0.2 mg per ml/
                 for **Euglena gracilis** : /at 1 mg per ml concentration/
   No Inactivating Effect on phage T4/N24/direct test/
                      on phage T4-multiplication in **E. coli**

Tests with f2-RNA Incorporated into Ointment

1. No resorption of P^{32}- labeled f2-RNA from rabbit skin/440 µg of f2-RNA per g/
2. No irritation of rabbit skin after 30 days of application/100 µg of f2-RNA per g/
3. No irritation of rabbit eye after 5 days of application/1 mg and 5 mg of f2-RNA tested/

[x] In parenthesis: results with "crude" preparation. The "purified" preparations of f2-RNA were obtained by employing the method of Kirby (1956) for removal of endotoxin.
[xx] The test was performed according to method of De Clercq et al./1973/

As the next step, the reactivity against f2-RNA was tested in
man. The results shown on Table 2 indicate the absence of dermal
hypersensitivity toward f2-RNA in 76 tested persons as well as
absence of hematological, liver and/or renal alterations in 20 dermatological patients treated with this preparation for a period up
to 30 days. In these cases, the toxicological monitoring was performed on admission and after termination of the topical therapy.

Table 2

Tests of Reactivity to f2-RNA in Humans

1. No signs of dermal hypersensitivity in 76 tested persons/
    epicutaneous application/. ,

2a. Pyrogenic reaction after intravenous administration of 1 mg
    f2-RNA to a cancer patient[x].

 b. No temperature rise after repeated intramuscular administration
    of 0.1-1.0 mg doses to five cancer patients. /Total: 2.85-10.6
    mg of f2-RNA/. /Slight lymphopenic effect in 2 cases and
    monocytosis observed in 3 cases/.

3a. No hematological, liver or renal alterations after 5 to 30
    days of local administration of f2-RNA containing ointment
    /1 mg per g/ to 20 dermatological patients.

 b. No hematological alterations in 2 patients with rabies receiving
    3 x 10 mg of f2-RNA intramuscularly on alternate days.

4. No hematological, liver or renal alterations after continuous
    application of f2-RNA ointment to the diseased early during
    a 2 years period/more than 500 mg of f2-RNA0.

[x] One of the first batches used for test-endotoxin effect?

No local or systemic disturbances were detected in a child with
an eye disease of neoplastic character/conjunctivitis lignosa/
treated for two years with a total amount of f2-RNA exceeding 500 mg
/28 mg per kg of body weight/. In 5 patients with advanced cancer
the intramuscular injections of f2-RNA in doses of 0.1 to 1 mg/
total 10.28 mg/were well tolerated. However, after intravenous
injection of 1 mg of f2-RNA a sharp temperature increase was
measured/Eckhardt-1973 person. communication/. Since the purification procedure of Kirby (15) was not applied to the batch used in
the latter case, the result gave no clear answer as to the systemic
tolerance of purified f2-RNA. Code (1977, person. communicatio)

recently treated intramuscularly two patients with rabies with 10 mg
doses of f2-RNA on alternate days for a week. No alteration of
hematological data was observed. Altogether, these results were
considered a sufficient safety basis for the subsequent topical use
of f2-RNA in dermatological and eye patients.

Clinical Trials with f2-RNA in Viral Skin and Eye Diseases

After the first therapeutic trials with f2-RNA in the Dermatological Clinic in Bratislava the tests were extended to 4 other
hospitals and an eye clinic. The ointment used in these tests
contained first 100 µg of f2-RNA per g of base. Later, when the
adverse effects seemed excluded, the amount of f2-RNA was enhanced
to 500 and 1000 µg per g.

Table 3 shows the array of skin diseases treated with f2-RNA
and the effectiveness of treatment expressed in the average duration of disease based on data reported by physicians. During the
past 5 years, more than 200 patients were treated with f2-RNA in
5 hospitals in Slovakia. In conclusion of this period the con-
sensus was reached that: a) the treatment with f2-RNA is harmless
for patients; b) causes, in the majority of cases, a rapid feeling
of relief, and c) shortens the duration of the disease.

In the same time it was learned that several factors may influence the effectiveness of f2-RNA therapy. One of them is an
underlying disease such as cancer which may cause a delay in the
healing effect (Table 3). An evident failure of f2-RNA requiring
a substitution therapy occurred in about 10% of herpes simplex
cases, 15% of herpes zoster cases and about 30% of verrucae of all
types treated with f2-RNA for a period shorter than 20 days. The
therapy with f2-RNA was ineffective also in 13 cases of non-viral
dermatoses supporting, although not proving, the inital assumption
that the curative effect was mediated through an antiviral mechanism.

Although a rapid healing, preceded by an antiinflammatory effect
and feeling of relief, was often reported by physicians employing
the f2-RNA therapy, the results on Table 3 do not allow a clear-cut
answer as to the therapeutic value of f2-RNA since the development
and termination of viral dermatoses shows great variations both
individually and during repeated attacks. The second factor complicating the assessment of the therapeutic value of f2-RNA is the
fact that the treated persons were, with exceptions of cases treated
in the Institute of Oncology, out-patients visiting the dermatological departments at various stages of the disease and with a difficult
control of the terapeutic regimen. To gain a better insight in this
problem, two approaches were utilized presently. First, patients
treated with f2-RNA for recurrent herpes simplex more than a year
ago were asked to express their individual view on the effectiveness

Table 3

Results of Treatment of Viral and Non-viral Dermatoses with f2-RNA Containing Ointment in 3 Hospitals

Disease	Dermatological Department	Number of Patients	Duration of the illness in days				Evident Failure
			without treatment^x	After treatment with f2-RNA			
				x̄	Sx	Sx̄	
Viral dermatoses							
1. Herpes simplex	Bratislava	64	5-14 days	3.814	1.308	0.170	
a) herpes labialis		31		3.476	1.249	0.272	
aa) herpes faciei				4.385	1.503	0.4168	4
2. Male herpes genit.	Bratislava	12	7-14 days	5.570	3.520	1.0161	1
3. Herpes zoster	Bratislava	17	14-21 days				3
	Kosice	12		7.313	2.358	0.589	-
	Inst. of Oncology	5		12.4			1
4. Verrucae vulg.	Bratislava	7	years(?)	17.42	3.5530	1.342	4
	Kosice	9					2
5. Condyloma acum.	Bratislava	2	years(?)				
Non-viral dermatoses							
1. Ulcus cruris	Bratislava	2					2
2. Bacterial eczema		5					5
3. Chronical eczema		2					2
4. Pemphigus vulg.		2					2
5. Glossitis erosiva	Kosice	2					2

x According to Jadassohn J.: Die Viruskrankheiten der Haut, Springer Berlin, 1961

of f2-RNA therapy through a questionnaire. It was hoped that this
approach may give an acceptable answer with regard to the frequency
of recurrences after therapy. Second, a collaborative double blind
experiment of limited extent was started in 3 hospitals to evaluate
the effectiveness of f2-RNA in new patients visiting the dermatological departments.

Table 4 shows that 79.3% of former patients consider the f2-RNA
therapy effective and 62.06% pf then consider it "better" than other
treatments. According to the view expressed in the questionnaire,
the f2-RNA therapy eliminated or substantially prolonged the intervals between herpes simplex attacks in 67.8% of former patients.
The fact that before the f2-RNA therapy 10 out of 29 respondents
suffered from repeated attacks monthly and additional 11 had attacks
bimonthly, lends certain credence to above data.

Table 4

Effectivity of f2-RNA Therapy of Herpes Simplex, Results of Evaluation of a Questionnaire[x]

1. The ointment with f2-RNA was effective :   79.3%
                                 ineffective :   20.6%

2. The effect was considered "excellent" :   24.13%
                               "very good" :   37.9%
                                   "good"   :   17.2%

3. After treatment, the disease:

    a) reappeared in previous intervals : 32.1%
    b) in prolonged intervals :   35.6%
    c) did not reappear :   32.1%
    d) no clear answer :   0.34%

4. The ointment with f2-RNA was considered

    a) better than other treatment :   62.06%
    b) worse than other treatment :   0.52%
    c) no clear answer :   37.9%

[x] Only patients with repeat attacks and treated with f2-RNA
    more than 1 year ago were evaluated. The results relate
    to 29 respondents.

Although the double blind study in the present form could not
eliminate the individual variations in the therapeutic regimen

leading to significant differences in the total amount of f2-RNA
applied to the diseased skin, a more than 90% effectiveness of
f2-RNA therapy in comparison with a 25% effectiveness of placebo
therapy was registered on a 10 days effect basis /P<0.001/. Less
significant was the difference on a less than 5 days basis suggesting a higher spontaneous healing tendency in this group /P<0.05/.
The results of testing and the symptomatological composition of
2 groups of patients under 15 years are shown on Table 5.

In contradistinction to herpes simplex and, in part, zoster
or male genital herpes, the therapeutic effect in verrucae vulgares
et planae can be assessed without difficulty be registering their
disappearance. The results of double blind experiment suggest a
70% therapeutic effectiveness of f2-RNA in warts but, due to the
small number of patients in tested groups, the statistical significance of results is presently inconclusive. During these studies,
it turned out that verrucae in patients under 15 years show a
faster healing/disappearing/tendency/average duration of treatment
13 days/than warts of adults/average duration of treatment 25 days/.
These results are in agreement with the calculations of effectiveness presented on Table 3. A prolonged treatment seems to be
required also in verrucae which proved resistent to such forms of
therapy as debridement or freezing.

Topical treatment of viral eye diseases with f2-RNA was up
to now limited to 4 cases of keratoconjunctivitis and 1 case of
conjunctivitis lignosa- a rare disease of suspected viral etiology with a sight threatening proliferation of conjunctival tissue
(18, 25). In the latter case it was encouraging to see that the
f2-RNA therapy led to a restitutio ad integrum without cicatri-
zation, and, despite of the significant total amount of f2-RNA
applied to the eye, caused no clinical and/or hematological
disturbances (Table 6).

The f2-RNA used in ophthalmological clinic was incorporated
in an ointment of ophthalmological formula /500-1000 µg of f2-RNA
per gram of base/. It exerted a rapid antiinflammatory effect
also in the therapy of more usual viral infections. Photograph 1
shows a 30 year old female with chronic keratoconjunctivitis
disciformis treated for weeks with mydriatics, Framycoin etc.,
without any effect. The patient came to the clinic with considerable alterations of cornea and visus. Shortly after starting
the f2-RNA therapy, the inflammatory signs disappeared and the
visus partially improved (Photograph 2). After about 10 weeks,
the healing of the epithelial cornea was complete. However, the
stromal alterations left a scar which requires a partial trans-
plantation. On the other hand, two acute cases of keratoconjunctivitis herpetica were cured completely in 14-18 days.

Table 5

Results of Treatment of Herpes Simplex with f2-RNA and/or Placebo

Patients treated with

Age Groups (in years)	Number of Cases	ds-RNA Cured (in days)				Number of Cases	Placebo Cured (in days)				Statistical Significance P:
		<5	>5 -<10	Total	In%		<5	>5 -<10	Total	In%	
<15	5^1	1	4	5	100	4^2	1	1	2	50	
>15 -<50	12	8	3	11	91.7	8	1	-	1	12.5	<0,001
Total	17	9	7	19	94.1	12	2	1	3	25	<0.001

[1]: Herpes simplex faciei permag. recid., 3 cases
Herpes simplex, empyema thoracis, 1 case
Herpes simplex in vulnero recid., 1 case

[2]: Herpes simplex labialis permag. recid., 1 case
Herpes simplex labialis recid., 1 case
Herpes simplex faciei recid., 1 case
Herpes simplex in vulnero, 1 case

Table 6

Hematological Findings in a Patient Treated with f2-RNA for 2 Years

	Erythrocytes x 10⁶	Leukocytes x 10³	Thrombocytes x 10³	Lymphocytes in %	Monocytes in %	Hemoglobin in %	F.W. 1 and hrs va
June 1, 1975	4.20	4.5	230	42	4	70	7/15
June 1, 1976	3.92	4.2	255	46	5	70	8/16
January 14, 1977	4.12	6.5	230	35	4	75	6/12
March 29, 1977	4.25	5.2	220	32	4	75	6/12

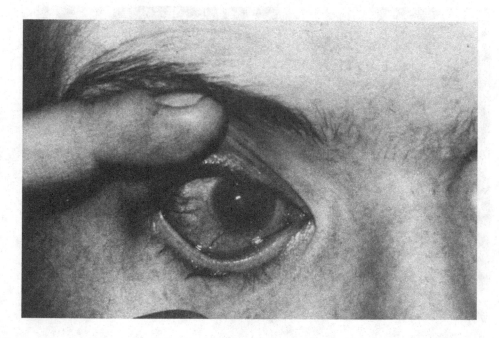

Figure 1

A 36 year old woman with keratoconjunctivitis disciformis chronica
before treatment with f2-RNA.

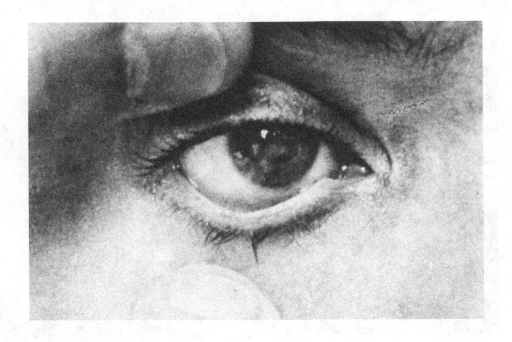

Figure 2

The patient shown on Photograph 1 after treatment with f2-RNA.

Discussion

The results obtained during five years of testing the topical therapeutical effectiveness of f2-RNA in more than 200 cases of viral dermatoses and a small number of viral eye diseases can be summarized as follows.

1. The clinical trials with f2-RNA eliminated the fears from the possible toxic consequences of topical RNA therapy in man. Although the absence of toxicity in patients with viral dermatoses can be, in part, explained by the low amounts of the drug required for achievement of the therapeutic effect /2.5 to 15 mg per person/, the maintenance therapy practicized in conjunctivitis lignosa, with a total of more than 500 mg of f2-RNA (28 mg per kg of body weight) applied to the diseased eye during a 2 year period, supports the view that the preparation represents no health hazard to man.

2. Due to the broad variation in severity and duration of herpes simplex both individually and by repeated attacks, the evaluation of effectiveness of f2-RNA therapy presents several problems. One is the fact that the majority of herpes simplex patients came to hospital with a developed disease while the f2-RNA therapy proved most effective when applied at the first day of symptoms. On the other hand, the high frequency of repeated attacks represents a certain unifying factor among patients visiting the hospital. For this reason, the effectiveness of a 10 day lasting therapy as basis for evaluation seems justified. Under such conditions, the f2-RNA therapy proved effective in more than 90% of cases while placebo was effective only in 25% of cases (P<0.001).

3. f2-RNA seems to be effective also in herpes zoster and male genital herpes. Both of these diseases may have serious sequelae. In addition, the former often complicates chronic systemic diseases or immunosuppressive therapy while the latter proved resistant to drugs like adenine arbinoside which showed promising efficacy in some studies (26, 27). This supports the perspective inclusion of f2-RNA into the therapeutic armament of dermatology.

4. The number of patients treated with f2-RNA in this study for viral eye diseases was insufficient for a statistical evaluation but the results warrant the perspectiveness of f2-RNA in ophthalmonology.

5. Of similar interest is the effectiveness of f2-RNA in therapy of warts and conjunctivitis lignosa since they represent diseases with neoplastic character.

6. According to some estimations, viral dermatoses represent about 12% of all skin diseases in childhood, and about 20% of adult population suffers from repeated attacks of herpes simplex

(28).   Some of viral dermatoses such as herpes zoster present therapeutical problems frequently complicating neoplastic diseases or
immunosuppressive therapy.   Finally viral eye diseases may have
debilitating consequences.   For all these reasons, search for drugs
with a broad spectrum of therapeutic activity is highly justified.
Although the results of this study require further specifications,
we believe that f2-RNA is one of candidates of such a drug because
it is cheap, relatively safe and easily manageable.

ACKNOWLEDGEMENT

    The authors thank Dr. J. Grunt for the statistical evaluation
of results.

REFERENCES

1. Braun, W., and M. Nakano. 1967. Antibody formation: Stimulation by polyadenylic and polycytidylic acid. Science (Washinton) 157:810-821.

2. Morrell, R.M. 1971. Effect of polynucleotides (poly I:C and poly A:U) on protein, RNA and polysomal synthesis in rabbit and human lung and rat and human peritoneal macrophages. Pgs. 179-195 in R.F. Beers, W. Braun eds. Springer Verlag Berlin-Heidelberg, New York.

3. Butlin, M., and P.G. Cunnington. 1976. Naturally-occurring double-stranded RNA and immune responses. IV. Influence of molecular size on antigenicity and adjuvant activity. Europ. Jour. Immunol. 6:607-612.

4. Serota, F.T., and R. Baserga. 1970. Polyinosinic acid-polycytidylic. Inhibition of DNA synthesis stimulated by Isoproterenol. Science (Washington). 167:1379-1380.

5. Hunter, T., T. Hunt, R.J. Jackson, et al. 1975. The characteristics of inhibition of protein synthesis by double-stranded ribonucleic acid in reticulocyte lysates. J. Biol. Chem. 250: 409-417.

6. Mécs, E. 1964. Inhibition of interferon action and production by adenine. Acta Virol. 8:475.

7. Levy, H.B., R. Adamson, P. Carbone, et al. 1971. Studies on the anti-tumor action of poly I:poly C. Pgs. 55-65 in R.F. Beer and W. Braun, eds. Biological Effects of Poly nucleotides. Springer Verlag Berlin-Heidelberg. New York.

8. Lacour, J. 1975. Trials with poly A:poly U as adjuvant therapy complementing surgery in randomized patients with breast cancer. Pgs. 229-232 in A.A. Gottlieb, O.J. Plescia, and D.H.L. Bishop, eds. In Fundamental Aspects of Neoplasia. Springer Verlag Berlin-Heidelberg, New York.

9. Field, A.K., A.A. Tytell, G.P. Lampson, et al. 1967. Inducers of interferon and host resistance. IV. Double stranded replications form RNA from E. coli infected with MS2 coli-phage. Proc. Nat. Ac. Sci. USA 58:2102-2108.

10. Banks, G.T., K.W. Buck, E.B. Chain, et al. 1970. Antiviral activity of double stranded RNA from a virus isolated from Aspergillus foetidus. Nature (London) 277:505-507.

11. Doskočil, J., N. Fuchsberger, J. Vetrák, et al. 1971.

Double-stranded f2-phage RNA as interferon inducer. Acta Virol. (Prague) 15:523.

12. Klienschmidt, W.J., J.L. Van Etten, A.K. Vidaver. 1974. In-fluence of molecular weights of bacteriophage phi 6 double-stranded ribonucleic acids on interferon induction. Infect. Immun. 10:284-285.

13. Billeter, M.A., and C. Weissman. 1960. Double stranded MS2-RNA from MS2-infected Escherichia coli. Pgs. 498-512 in G.L. Cantoni and D.R. Davies. Procedures in Nucleic Acid Research. Harper and Row, New York and London.

14. Franklin, R.M. 1966. Purification and properties of the replicative intermediated of the RNA bacteriophage Rl7. Proc. Natl. Acad. Sci. (Washington) 55:1504-1511.

15. Kirby, K.S. 1956. A new method for the isolation of ribonu-cleic acids from mammalian tissue. Biochem. Jour. 64:405-408.

16. Peakock, R.D., and C.W. Dingman. 1967. Molecular weight estimation and separation of ribonucleic acid by electro-phoresis in agarose-acrylamide composite gels. Biochemistry 7:668-674.

17. Palaček, E. and J. Doskocil. 1974. Pulse-polarographic analysis of double-stranded RNA. Analyt. Biochem. 60:518-530.

18. Borecký, L., J. Doskočil, V. Lackovič, et al. 1975. Double stranded RNA of Czechoslovak production in the therapy of viral skin- and eye-diseases. Pgs. 149-158 in D. Ikic' and Zagteb eds. Proc. Symposium on Clinical Use of Interferon.

19. Mayer, V., E. Mitrová, E. Gajdošová, et al. 1974. Viral infection and resistance in immunosuppressed host. V. Con-version of enhanced fatal tick-borne encephalitis to a non-lethal form by a resistance stimulator of the double-stranded RNA type. Acta. Virol. (Prague) 18:31-41.

20. Gajdošová, E., V. Mayer, and J. Doskočil, 1973. Viral in-fection and resistance in immunosuppressed host. IV. Abo-lition of potentiated togavirus A infection by a resistance inducer of the double-stranded RNA type. Failure of an exogenous interferon preparation. Acta Virol. (Prague) 17:327-337.

21. Stinebring, W., and M. Absher. 1971. The double-bitted axe: A study of toxicity of interferon releasers. Pgs. 249-257 in R.F. Beers and W. Braun. Springer Verlag-Berlin-Heidelberg-N. York.

22. De Clercq, E., W.E. Stewart II, and P. De Somer. 1973. In-
     creased toxicity of double stranded ribonucleic acid in virus
     infected animals. Infect. Immun. 7:167-172.

23. Táborský, I., J. Doskočil, and V. Zajicova. 1974. Pyrogenic
     activity of double-stranded RNA from Escherichia coli phage
     f2. Acta Virol. (Prague) 18:106-112.

24. Rovenský, J., J. Doskočil, J. Pekárek, et al. 1975. Pre-
     vention of spontaneous autoimmunity to DNA in NZB/Swiss mice
     by treatment with natural ds-RNA. Immunology 29:745-748.

25. Obručníková, E., V. Svec, and L. Borecky. 1977. On diagnosis
     and therapy of conjunctivitis lignosa (in Czech) Cs. ophthal-
     mologie. In press.

26. Whitley, R.J., L.T. Chien, R. Dolin, et al. 1976. Adenine
     arabinoside therapy of herpes zoster in the immunosuppressed.
     New Engl. J. Med. 294:1193-1199.

27. Adams, H.G., L.A. Vontver, E.A. Benson, et al. 1976. Genital
     herpetic infection in man and women: Clinical course and
     effect of topical applicaiton of adenine arabinoside. J.
     Infect. Dis. Suppl. 133:151-159.

28. Borzow, M.V., V.P. Kuznecow, and G.I. Lobanovskiy. 1971.
     Data to application of interferon in the therapy and pro-
     phylaxis of viral dermatoses (in Russian). Vestn. Derm.
     Vener. 45:14-17.

THE POSSIBILITY OF INTERFERON PRODUCTION IN TUMORS

Warren R. Stinebring and Stephen Jenkins

University of Vermont
Department of Medical Microbiology
College of Medicine
Burlington, Vermont   05401

Endotoxin* injected into tumor bearing animals and humans may result in regression or even elimination of the tumor(1,2). Endotoxin also causes release of host products, including interferon(3), an antiviral factor which also has effects on tumor cell replication in vitro and in vivo in animals and possible humans(4), and tumor necrosing factor (TNF), which may be the basis for direct injury to the tumor(5).

Production of interferon in the animal is best considered as a local phenomenon. Evidence indicates that cells only produce interferon in response to a direct stimulus provided by virus, bacteria, or other materials. Thus, there seems to be no secondary signal, at least as far as is known, which results in a generalized production of interferon by cells not directly affected by the initial stimulus itself. The usual route of injection of inducers is either intravenous or intraperitoneal in order to obtain detectable titers of interferon in the circulation. The reason that the circulation is studied is obvious - the ease of sampling of blood. Even the production of interferon here is a local phenomenon. The reticuloendothelial system's phagocytic cells are responsible for the release of interferon into the circulation. These target cells are good producers of interferon, good exporters of interferon, and

* The term endotoxin will be used for substances derived from Gram negative bacteria from which injured tumors regardless of whether technically these substances are phospholipoprotein complexes, lipopolysaccharides, lipid A or ill-defined materials.

of course, the fluid they export it into is the blood. Under
these circumstances, immediate dilution of the interferon with
the subsequent lowering of concentration is apparent.

In some unique and very interesting experiments, Stevens and
Merigan(6) examined the titers of interferon attained in varicella-
zoster vesicles of human patients. They noted that interferon was
rarely demonstrable in the serum of patients and then only at low
levels of 10 to 30 units per 4 ml for a day or so. However, in
the vesicles, titers of 5,000 to 35,000 units were observed indicating that the local concentration of interferon was far higher than
that attained in general circulation. The authors remarked that the
lesion declined approximately 48 hours after the peak titers had
been reached in the vesicle fluids. Although the total amount of
interferon produced in these vesicles must be very considerable, it
is the local concentration which seems to be most important.

Experimental work using monolayers of susceptible cells in the
usual assay systems has also indicated that it is the concentration
of interferon rather than the absolute amounts available which is
important in antiviral activity. The work of Hallu, Thacore and
Youngner(7), clearly indicates that exogenous interferon used to
treat L-cells persistently infected with Newcastle's disease virus
(NDV), had to be present at 200 to 1000 units per culture, if the
number of infectious centers was to decrease. Continuous presence
of 50 units of exogenous interferon in the medium had no influence
on the infectious center counts even when the interferon was present
for as long as 78 days duration throughout, of course, a number of
cell cultures. 1,000 units of exogenous interferon resulted in a
more rapid elimination of infectious centers, so that by 14 days
infectious centers had dropped to practically nil. The authors
state that the cured cells, unlike the parent persistently infected
cells, were susceptible to vesicular stomatitis virus(VSV) infection
and did not revert to the persistently infected state.

One could also reasonably expect to see concentration problems
arise in the treatment of tumors by interferon therapy. Unless
there is some concentrating effect operative at the local site, one
might not expect to reach levels of interferon necessary in many
cases, for amelioration of the disease through interferon's effects
on cells. The local production of high concentrations of interferon
would seem to be a better therapeutic regimen than the broadcast
introduction of interferon at high dosage in such a manner as to
produce low local concentrations. This is especially true if one
wishes to affect metastases which may not be readily observable.
Something to seek these out, and we feel endotoxin may be one of
the materials, would be particularly useful expecially if such a
substance would produce high local concentrations of anti-tumor
substances as well as, perhaps, injure the vasculature of the tumor.

The rationale behind the experiments reported herein was
that it might be possible to produce high local levels of interferon
in tumors injured by endotoxin. Fisher rats were inoculated with
hepatomas. Animals used in this experiment were randomized by use
of a table of random numbers. The hepatoma F4 (Obtained from P.
Kelleher and C. Smith, University of Vermont), was injected in the
flank in cell numbers expected to generate tumors in 100% of the
animals. Animals were fed rat pellets ad libitum, and maintained
until the tumors had reached the size desired, approximately 17-
19 days. Animals were then injected with Serratia marcescens lipo-
polysaccharide (Difco Laboratories), in a dose of 3000 micrograms
administered intravenously. Diluent was non-pyrogenic saline. Three
hours later the rats were killed, serum samples were obtained, spleens
were removed and tumors were excised. Studies on interferon content
were done on dilutions of the serum as obtained. 10% extracts of
the spleen were produced by grinding the spleen with sand in a mortar
and pestle and adding 9 times the weight of the spleen of tissue
culture media. The solid tumor tissue was treated as spleen, whereas
liquid which accumulates in large quantities in these tumors was
treated as was the serum. Control tumor bearing animals received
non-pyrogenic saline in the same volume used in the experimental
animals. Control animals without tumors were treated either with
saline or with endotoxin. Rat embryo fibroblasts of various passages
maintained in Linbro plates with 16 mm flat bottom wells were used
for interferon assay. Dilutions of fluids in tissue culture medium
were placed on the cells and allowed to remain overnight. The cells
were challenged with vesicular stomatitis virus(VSV) using 100
tissue culture infectious doses per well. Controls for virus
activity were set up at the same time. Cultures were read for
cytopathic effect on days 2, 3, and 4, following inoculation with
virus. Cytopathogenic effect is scored, ranging from 4+ which was
complete destruction of the monolayers to 0 which was no effect
at all. The end point of titration was taken as that dilution of
interferon which protected 50% of the cells from distruction by
VSV. Controls of virus activity showed 100% destruction. Titers
are reported as a reciprocal of the dilution protecting 50% of the
cells.

Table 1 indicates the results of the experiment. These results,
in general, confirm those of a previous experiment in which rat
hepatoma F3 had been used and interferon produced in detectable
quantities in tumor extracts of these tumors following challenge
with endotoxin. Note should be made that in 6 control tumor
bearers which received saline there was evidence of some minimal
(1:40) antiviral activity in the tumor fluid in two. The reason
for this is not known, although it is possible that infection
of the tumor or animal could account for these results. Saline
injected control normal rats showed no interferon activity. That
the antiviral material from the experimental animals receiving

TABLE I

Interferon-Like Effect Using Various Specimens From Rats

	<40	40	80	Titres - No. of Animals 160	320	640	1280	Animals, Total
Tumor bearing								
Serum	1	4	1	3	2	1	0	12
Spleen	x	x	x	2	4	3	0	11
Tumor Fluid	x	x	x	x	0	0	11	11
Normal								
Serum	1	1	3	3	1	0	0	9
Spleen	x	x	x	1	1	3	3	8

All animals received 3.0 mg S. marcescens endotoxin.

x = toxic for cells; indicates level of sensitivity.

endotoxin was likely to be interferon was shown by inability of
washing to remove protective action from the cells, by failure to
demonstrate neutralizing activity against the virus directly, and
by determining that this antiviral substance was not active on human
cells.*

Our interpretation is that, in deed, we may be able to injure
or affect the normal cellular components of the tumor in such a
fashion that the vascular system of the tumor is disrupted, and
that certain specific products such as interferon are produced
which, in turn, may inhibit the tumor. Controlling the cellular
components of the tumor, expecially wandering cells such as
lymphocytes and macrophages, may allow one to enhance local production of cell mediators effective against tumor cells. At the same
time, other normal tissues or normal cells of the host may not be
severely injured or injury may be of lesser consequence, e.g.,
cells of the reticuloedothelial system which are known to react
to endotoxin and release interferon as well as other substances, can
be sacrificed without permanent deleterious effects to the host.

* Subsequent experimentation has shown the antiviral effects to
be due to substances which have the characteristics of fibroblast
and leukocyte interferon (to be published).

We feel it is important that studies of passively administered
cell mediators such as interferon or possible tumor necrosing factor
be appropriate to study the production of these substances in high
local concentrations in the tumor in situ.

REFERENCES

1. Gullino, P.M. 1966. The internal milieu of tumors. Prog. Exp. Tumor Res. 8:1-25.

2. Folkman, Judah. 1974. Tumor angiogenesis factor. Can. Res. 34:2109.

3. Nauts, H.C. et al. 1946. The treatment of malignant tumors by bacterial toxins as developed by the late William B. Coley, M.D., reviewed in the light of modern research. Can. Res. 6:205.

4. Nauts, H.C. 1973. Enhancement of natural resistance to renal cancer: Beneficial effects of concurrent infections and immunotherapy with bacterial vaccines. N.Y. Can. Res. Inst. Monograph #12.

5. Anon. 1934. Erysipelas and prodigiosus toxins (Coley). Reports of the Council. Council on Pharmacy and Chemistry. J.A.M.A. 103:1067.

6. Stinebring, W.R. and D. Stevens. 1977. Escherichia coli endotoxin effect on a methylcholanthrene-induced sarcoma in the hamster cheekpouch. Proc. Soc. Exptl. Biol. Med. 156:229.

7. Evans, H.M. 1908. On the occurrence of newly-formed lymphatic vessels in malignant growths. Johns Hopkins Hospital Bulletin 209:232.

8. Zeidman, I. et al. 1955. Experimental studies on the spread of cancer in the lymphatic supply in carcinoma. Cancer 8:123.

9. Alexander, P. et al. 1976. The significance of macrophages in human and experimental tumors. Ann. N.Y. Acad. Sci. 276:124.

10. Stinebring, W.R. and J.S. Youngner. 1964. Patterns of interferon appearance in mice injected with bacteria or bacterial endotoxin. Nature 204:712.

11. Carswell, E.A. et al. 1975. An endotoxin-induced serum factor that causes necrosis of tumors. Proc. Nat. Acad. Sc. (USA) 72:3666.

12. Stevens, D.A. and T.C. Merigan. 1972. Interferon, antibody, and other host factors in herpes zoster. J. Clin. Invest. 51: 1170.

13. Hallum, J.V., H.R. Thacore and J.S. Youngner. 1972.
     Effect of exogenous interferon and L cells persistently
     infected with Newcastle Disease Virus. Infect. Imm.
     5:145.

14. See this volume.

THE FUTURE OF INTERFERON AS AN ANTIVIRAL DRUG

J.K. Dunnick, Ph.D. and G.J. Galasso, Ph.D.

National Institute of Allergy and Infectious Diseases

National Institutes of Health, Bethesda, Maryland   20014

INTRODUCTION

One of the exciting aspects of the twenty year "interferon era" is that research in this field has been characterized by international collaboration and information exchange. A 1957 report in the Proceedings of the Royal Society by Isaacs, of England, and Lindenmann, a fellow in his laboratory from Switzerland, ushered in the interferon era (1). In retrospect, Ørskov and Andersen in 1938 probably also observed interferon production after vaccinia infection of rabbit skin (2). It is evident from a review published in 1963 that by then workers from all over the world understood the potential of interferon and were characterizing the parameters of this drug(3).

The biologic activity of interferon is species specific (4); for exogenous interferon to be active in man it must be produced from human cells. Initial work on the purification of interferon (5,6) was begun shortly after its discovery, and though interferon can now be purified 5000 fold, absolute purification has not been achieved. During the 1960's, it was recognized that repeated doses of interferon had to be given to maintain therapeutic levels of exogenous interferon in man. Thus, to treat one patient requires large amounts of exogenous interferon. To obtain enough interferon to conduct clinical trials it is important to develop technology to produce and purify interferon.

The development of interferon reference reagents, including interferon standards and antisera, enabled results from one laboratory to be compared to results from another. In the mid

1960's interferon inducers were recognized, but due to inactivation
by hydrolytic enzymes in normal serum, inducers such as poly-
riboinosinic-polyribocytidylic acid were not optimal antiviral
agents in man (7). Recent work with the inducer polyriboinosinic-
polyribocytidylic acid-poly-1-lysine (8), suggests that this
compound can induce fairly high levels of interferon in man
(H. Levy, Personal Communication). Borecky and co-workers (9)
report that they have used another interferon inducer,
"Czechoslovak Double-Stranded RNA", to treat viral dermatoses in
man.

    Recent and ongoing clinical trials using exogenous interferon
are summarized in Table 1. Many of these studies have demonstrated
the clinical potential of human leucocyte interferon through open
trials, and double-blind controlled trials are now being conducted
in the treatment of many diseases. The dose and route of
administration of human leucocyte interferon vary with the disease
to be treated. When one is treating acute diseases, such as
cytomegalovirus or herpes zoster infections, relatively high amounts
of interferon are needed. In chronic infections, such as chronic
hepatitis B where there is a reduced viral load, interferon is
administered over extended periods and the effective daily dose
can be reduced 10-100 fold. The chosen routes for treating many
types of systemic diseases have been by intramuscular or subcutaneou
inoculation, and the reasons for choosing these routes of administration are twofold: 1) after intramuscular injection, interferon
is absorbed more slowly and circulates for longer periods of time
than after intravenous injection, and 2) patients who must receive
daily injections can administer the drug themselves by these
routes. For treatment of viral diseases of the eye interferon
is applied topically. Intranasal application of interferon at
high doses appears to be effective in treating some respiratory
illnesses, although much of the interferon is removed by natural
nasal clearance factors. Better ways to administer interferon
need to be developed so as to reach the target organ more
effectively and prevent the destruction or removal of interferon
by normal clearance mechanisms.

    Interferon has been shown to hold promise for treating RNA
viruses in man, including influenza, rhinoviruses, and rubella;
for treating DNA viral infections, including hepatitis B, cytomegalovirus, herpes, and varicella-zoster; and also for treating
certain metastatic diseases such as Hodgkin's disease and osteosarcoma. Side effects from interferon treatment have been
relatively minor; some side effects are fever, nausea and vomiting,
malaise and muscle aches. With continued high doses of interferon
a depression of white cells may be seen. These initial clinical
trials indicate that interferon can be used to treat a variety
of infectious diseases. Other infectious diseases with potential

for therapeutic application of human interferon include rabies, arbovirus infections, arenavirus infections (Lassa fever) and slow virus infections.

As we plan for the future, it is readily apparent that we must produce interferon in large quantities and learn how to purify it by removing unwanted contaminants. Currently there are 3 types of interferons available: leucocyte, fibroblast, and lymphoblastoid interferon. A certain amount of foresight is needed before large quantities of the latter two types will be stockpiled; one needs to know that the product is clinically useful and also meets the requirements of licensure in a particular country, for example by the FDA in the U.S. Lymphoblastoid interferon is being produced from cells which carry a portion of the EB virus genome, and its acceptability for treatment of infectious diseases needs to be determined. Fibroblast interferon can be produced from carefully karyotyped cells which meet the FDA requirements; the cells can be frozen and distributed throughout the world for use as interferon producers. Preliminary work using interferon prepared from human fibroblast interferon suggests that this type of interferon may also be an effective treatment against hepatitis or vaccinia infections. Comparison of the effects of human leucocyte interferon and human fibroblast interferon in animals and humans will help to determine the relative merits of these preparations. Human leucocyte interferon has been tested in a variety of diseases, and is the one type of interferon which has shown clinical efficacy. However, to produce this interferon a pool of leucocytes from thousands of individuals must be used, and these cells might also contain adventitious agents. New techniques such as inserting the genes for interferon production into bacteria might offer a new way to produce interferon economically.

Techniques for purifying interferon are now becoming available. Through column chromatographic techniques interferon can be purified 5000 fold with a specific activity of greater than 1×10^8 units/mg protein. Affinity chromatographic procedures using specific interferon antisera might offer a way of purifying interferon to an even greater extent. When using column chromatography to purify interferon, one must be sure that unwanted column materials such as concanavalin A do not contaminate the final product. Large scale production facilities are available for preparing lymphoblastoid interferon, and it is likely that fibroblast interferon could also be produced in large quantities. Recent work on membrane receptors for interferon suggests that specific gangliosides in the cell membrane are involved in binding interferon to the cell. This knowledge could potentially be used in purifying interferon; interferon could first be bound to

TABLE I

Clinical Administration of Exogenous Interferon

STUDY	DRUG	RESULTS	INVESTIGATOR
Herpes Viruses			
Herpes zoster and varicella zoster in patients with malignancy	HLI – 5.1×10^5 u/kg/day	some clinical improvement	Jordan et al[10] Mergian* Stanford University
Herpes Keratitis	HLI – drops (1.4×10^6 u/ml)	ongoing	Kaufman* University of Florida
Herpes keratitis	HLI – drops (3×10^6 u/ml)	some clinical improvement	Sundmachen[11] University Eye Clinic Freiburg, Germany
Herpes simplex following surgery for Tic Douloueaux	HLI – 2.5×10^7 u every 12 hours for 5 days	ongoing	Ho* University of Pittsburgh
Herpetic Keratitis	HLI – 11-31×10^6 u/ml – double blind study	some clinical improvement	Jones et al[12] Mooresfield Eye Hospital, England
Cytomegalovirus	HLI – 1.7-3.5×10^5 u/kg for 7-14 days	no definite conclusions can be drawn from trail	Arvin et al[13] Stanford University

STUDY	DRUG	RESULTS	INVESTIGATOR
Cytomegalovirus	HLI – $1-5 \times 10^6$ u for 8-10 days	some signs of clinical improvements	Emodi[14,15] Children's Hospital Basel, Switzerland
Cytomegalovirus after bone marrow transplantation	HLI – $2 \times 10^5 - 1 \times 10^6$ u/day for 5-10 days	some clinical improvement	O'Reilly et al[16] Memorial Sloan-Kettering Cancer Center
Myxoviruses			
Hong Kong Influenza	HLI – repeated intransal application of interferon (1000 u/ml)	some clinical improvement	Solov'ev[17] Academy of American Sciences – USSR
Influenza B	HLI – total dose 800,000 u intransal	no effect	Merigan[18] Clinical Research Centre, England
Picornoviruses			
Rhinovirus	HLI – total dose 14,000,000 u given intranasally	some clinical improvement	Merigan et al[18] Clinical Research Centre, England
Rubella			
Congenital rubella	HLI – 3×10^6 u/day for 14 days	clinical improvement	Larsson et al[19] Karolinska Institute

STUDY	DRUG	RESULTS	INVESTIGATOR
Hepatitis			
Chronic hepatitis B	HLI - 6.0×10^3 - 1.7×10^4 u/kg for several months	clinical improvement	Greenberg et al[20] Merigan* Stanford Hospital
Chronic hepatitis B	HFI - 10^7 u every other day for 14 days	some clinical improvement	Desmyter[21] Rega Institute
Acute hepatitis B	HFI	ongoing	Rega Institute*
Chronic active hepatitis B	HLI	some clinical improvement	Zuckerman* London School of Hygiene and Tropical Medicine
Pox Viruses Vaccinia Vaccination	Rhesus monkey interferon	produced protection	Isaacs et al[22] Scientific Committee on Interferon,England
Vaccinia	HFI - 10^5 u at site of vaccination	reduced % of vaccination takes	Tyrrell* Clinical Research Centre, England
Metastatic Diseases Hodgkin's disease	HLI - 7 months treatment - total dose 1377 million units	some remission of disease	Blomgren et al[23] Karolinska Institute

STUDY	DRUG	RESULTS	INVESTIGATOR
Osteosarcoma	HLI - treatment for 1 year	60% have no metastasis after surgery	Strander et al24 Karolinska Institute
Breast Cancer	HLI	pilot study	Habef* Columbia University
Adenocarcinoma of the lung	HLI - pilot studies with 2 patients - 3x10⁶ u two times a week	feasibility study - no conclusion can be drawn on prevention of metastasis	Oettgen* Memorial Sloan-Kettering Cancer Center
Non-Hodgkin's lymphoma	HLI - 10x10⁶ u/day for 30 days	no striking effect on measurable disease in 3 patients - ongoing trials to be extended to other tumors	Merigan and Rosenberg* Stanford U. Hospital
Juvenile larynx papilloma	HLI - 3x10⁶ u three times a week	some clinical improve-ment	Strander* Karolinska Institute
Multiple myeloma	HLI - 3x10⁶ u twice daily	pilot study	Strander* Karolinska Institute

STUDY	DRUG	RESULTS	INVESTIGATOR
Reduction in virus infection after transplantation			
Renal transplant	HLI - 6 week treatment	ongoing	Hirsch* Massachusetts General Hospital
Renal transplant	HFI - 6 week treatment	ongoing	Schellekins* Erasmus Hospital Holland
Miscellaneous			
Influenza, Herpes, tumors and cervical cancers	HLI	some clinical improvement	Ikic et al* Institute Immunology Zagreb, Yugoslavia

HLI = human leukocyte interferon

HFI = human fibroblast interferon

u = units of interferon

* = personal communication

specific receptors to separate it from the milieu of the cell supernatant; the interferon could then be released from the receptor to
give a purified product. During the various steps involved in
the production and purification of interferon, and in monitoring
interferon activity <u>in vivo</u>, one needs to assay its activity.
The development of rapid assay systems, such as radioimmunoassays,
would greatly increase the speed and accuracy at which these
procedures could be accomplished.

Success in conducting future clinical trials in man depends
on the continued cooperation of scientists in this field including
people from government, industry, and the medical community at
large. Rapid exchange of information has enabled scientists
throughout the world to gain knowledge on appropriate doses and
routes of administration, and it is likely that continued
information exchange will be one of the primary reasons for the
successful development of interferon as an antiviral for use in
man. Recognizing the potential of this drug, government and
industry must continue to support efforts to purify and produce
enough interferon to conduct future clinical trials. The promise
of interferon lies in the fact that it has been found to be a
broad spectrum antiviral active against many viral diseases
in man; it has low toxicity and a high specific activity. It
can be expected that interferon will eventually be used for
treating certain infectious diseases and metastatic diseases,
as a supplement to be given with vaccines, and as a prophylactic
or therapeutic drug in cases when vaccine treatment has not
been given, or where vaccines are not available.

REFERENCES

1. Isaacs, A., J. Lindenmann. 1957. Proc. Roy. Soc.
    Series B 147:258-267.

2. Ørskov, V.J., E.K. Andersen. 1938. Weitere Untersuchungen
    über die Bildungsstätten der virusneutralisierenden Stoffe
    bei Vaccineinfektion von Kaninchen. Acta Pathol. Microbiol.
    Scand. Suppl. 37:621-631.

3. Isaacs, A. 1963. Interferon. Pgs. 1-38 in K.M. Smith,
    M.A. Lauffer, eds. Advances in virus research volume 10.
    New York and London, Academic Press.

4. Tyrrell, D.A.J. 1959. Interferon produced by cultures of
    calf kidney cells. Nature 184:452-453.

5. Burke, D.C. 1961. The purification of interferon. Biochem.
    J. 78:553-556.

6. Lampson, G.P., A.A. Tytell and M.M. Nemes, et al. 1963.
    Purification and characterization of chick embryo interferon.
    Proc. Soc. Expt. Biol. Med. 112:468-478.

7. Field, A.K., A.A. Tytell and G.P. Lampson, et. al. 1967.
    Inducers of interferon and host resistance. Multistranded
    synthetic polynucleotide complexes. Proc. Natl. Acad.
    Sci. 58:1004-1010.

8. Levy, H.B, G. Baer, and S. Baron, et al. 1975. A modified
    polyriboinosinic-polyribocytidylic acid complex that
    induces interferon in primates. J. Infect Dis. 132:434-439.

9. Borecky, L., J. Doskocil and V. Lackovic, et. al. 1975.
    Double-stranded RNA of Czechoslovak production in the therapy
    of viral skin and eye diseases. Proc. Symposium on Clinical
    Use of Interferon. 149-158.

10. Jordan, G.W., R.P. Fried and T.C. Merigan. 1974. Administra-
    tion of human leucocyte interferon in herpes zoster. J.
    Infect Dis. 130:56-62

11. Sundmachen, R., D. Neumann-Haefelin and K. Cantell. 1976.
    Successful treatment of dendritic keratitis with human
    leukocyte interferon. Albrecht v. Graefes Arch Klin Exp.
    Ophthal. 201:39-45.

12. Jones, B.R., D.J. Coster and M.G. Falcon, et al. 1976.
     Topical therapy of ulcerative herpetic keratitis with human
     interferon. Lancet 2:128.

13. Arvin, A.M. A.S. Yeager and T.C. Merigan. 1976. Effect
     of leukocyte interferon on urinary excretion of cytomegalovirus
     by infants. J. Infect Dis. 133 (Suppl.):A205-210.

14. Emodi, G. and M. Just. 1974. Impaired interferon responses
     of children with congenital CMV disease. Acta. Pediatric
     Scand. 63:183-187.

15. Emodi, G., R. O'Reilly and A. Muller, et al. 1976. Effect
     of human exogenous leukocyte interferon in cytomegalovirus
     infections. J. Infect Dis. 133 (Suppl):A199-204.

16. O'Reilly, R.J. L.K. Everson and G. Emodi, et al. 1976.
     Effects of exogenous interferon in cytomegalovirus infections
     complicating bone marrow transplantation. Clin Imm.
     Immunopathol. 6:51-61.

17. Solov'ev, V.D. 1969. The results of controlled observations
     on the prophylaxis of influenza with interferon. Bull Wld.
     Hlth. Org. 41:683-688.

18. Merigan, T.C. S.E. Reed and T.S. Hall, et al. 1973.
     Inhibition of respiratory virus infection by locally applied
     interferon. Lancet 1:563-567.

19. Larsson, A. M. Forsgren and S. Hard, et al. 1976.
     Administration of interferon to an infant with congenital
     rubella syndrome involving persistent viremia and cutaneous
     vasculitis. Acta Paediatr Scand 65:105-110.

20. Greenberg, H.B., R.B. Pollard and L.I. Lutwick, et al. 1976.
     Effect of human leucocyte interferon on hepatitis B virus
     infection in patients with chronic active hepatitis. N.
     Engl. J. Med. 295:517-522.

21. Desmyter, J. M.B. Ray and A.F. DeGroote, et al. 1976.
     Administration of human fibroblast interferon in chronic
     hepatitis B infections. Lancet 2:645-647.

CONSIDERATIONS IN OUR SEARCH FOR INTERFERONS FOR CLINICAL USE

R.Z. Lockart, Jr., Ph.D. and E. Knight, Ph.D.

Central Research and Development Department
E.I. du Pont de Nemours and Company
Experimental Station
Wilmington, DE 19898

ABSTRACT

Interferon has yet to become the drug of choice for any
clinical disease entity, but several promising uses are being
pursued. Clinical trials are limited probably because
interferon is so expensive. Cheaper interferon would probably
stimulate more clinical trials. Demonstrated efficacy in a
disease of considerable magnitude would spur the search for
cheaper interferon. Possible ways to look for more, cheaper
interferon are suggested. The need to understand the chemical
and structural composition of interferon is pointed out especially
the role played by carbohydrates. There is a need for a non-
biological, yet sensitive quantitative measurement of interferon;
development of an immune assay is suggested. The need to restudy
ways to apply interferon to target tissues and the need for a
standardized product for use in clinical trials are pointed out.

INTRODUCTION

Interferon fortunes have fluctuated over the years like a
giant sine wave. From the time of its discovery it has enjoyed
considerable appeal as the major answer to, the ideal way to
prevent and treat, viral diseases. There is little doubt that
interferon plays an important role as an early defense mechanism
in the body's fight against invading viruses and perhaps other
foreign organisms. Administered interferon has been shown to
prevent viral infections of laboratory animals and man.

Why then hasn't interferon found greater use? Let me present a few ideas. Interferon has limitations. It has not proved particularly useful in acute infections when given after symptoms appear. It has to be injected. When given prophylactically, it gives only temporary protection, thus to maintain a protected state, repeated injections must be given. Vaccines have been developed against most of the viral diseases whose acute infections occur in epidemics or which are life-threatening in their severity. Those virus diseases which are in the above category but which cannot yet be controlled by vaccines, such as influenza, have likewise not given satisfactory indications of being controllable by interferon. Interferon for clinical use has therefore seemingly been relegated to more chronic conditions. It is expensive.

As I see it, the major need of interferon from the standpoint of clinical use in humans, is the demonstrated efficacy of interferon in the prevention or treatment of a disease with a sufficient number of annual cases so that a market is created to justify its expense. It must become the treatment of choice for some disease.

I have heard the question, "Where are we going to get enough interferon to do the clinical trials we need?" From the reports I heard on interferon production at a recent workshop held at The National Institutes of Health, it appears that there now exists at a minimum, the capability of producing 1×10^{10} units per year of leukocyte interferon and approximately 3×10^{10} units per year of human diploid fibroblast interferon. I wonder if the question should be, "Who is going to purchase the interferon and finance the trials?" The two diseases of considerable magnitude for which there have been encouraging clinical results are osteogenic sarcoma and hepatitis B infection, both of which have been discussed in this meeting. Each requires long-term introduction of interferon. At the doses of interferon given which produced the results, estimating interferon at a cost of $100 per 1×10^6 units, it costs about $4000 per month per patient for interferon alone to do clinical trials and it takes a year or two to get sufficient data for evaluation. In the case of hepatitis B chronic carriers, with daily doses of $0.5-2.5 \times 10^6$ units, it will cost $1,500 to $7,500 per month per patient or an annual cost of $18,000 to $90,000 for the 180 to 900×10^6 units per patient for such a study. To do a modest clinical study, of say 20 such patients, $360,000 to $1.8 million worth of interferon alone is required, but it can be done with less than 2×10^{10} units, amounts already available. Herein lies the catch-22 with interferon and disease. There is a need for cheaper interferon so that more clinical testing can occur and there is a need for a demonstrated use of interferon to spur the search for cheaper interferon. A significant

advance in either will surely provide an impetus for the other.
Hopefully, one or the other will occur before enthusiasm for
the clinical use of interferon dies.

Since production of more, cheaper interferon is a major
need, I would like to direct a few remarks to that subject.
Already, in several laboratories, interferon has been produced
in large scale fermenters. The major problem with the interferons from these sources at present is that they come from
cell lines of lymphoid origin isolated from people suffering
from Burkitt's lymphoma and must either be given as a very
purified product or one amenable to quality control and extensive
safety testing. Research is needed, therefore, to find
acceptable cells which can make interferon in fermentation tanks
and to work out the conditions for maximal interferon production.
I am not aware of any work directed toward this end. There is
even very little work looking at new cell sources for interferon
production and this might be more imaginatively attacked. One
possibility that would get around some or many of the objections
aimed at the products of human cells is the possibility that
interferon from animal cells other than human might serve as
a source of interferon. Clearly, interferons from mice, chickens,
and rabbits seem to be excluded from human use, but interferons
from pigs or cows might work.

Another line of research that I have not yet seen pursued
much that might offer new insights into the specificity,
stability, and pharmaco-kinetics of interferons is one which
would seek to understand the role of the carbohydrate moiety
of the interferons. Both receptor specificity, and clearance
rates of the glycoprotein enzyme asparaginase have been changed
by alterations in the carbohydrate portions of the enzyme
molecules. A single sugar added to some of the blood group
glycoproteins changes completely their antigenicity. Surely,
there is growing evidence that the carbohydrate portion of a
number of molecules may play an important role in its biological
action. Changes in the carbohydrate of interferons could
conceivably alter their specificities, their stabilities, and
their rates of clearance when injected. But it is obvious
that we not only don't know much about the carbohydrate portion
of interferon, we know very little about their other chemical
characteristics. Knowledge of the amino acid sequence might
lead to possible synthetic routes of either the whole molecule
or perhaps active fragments.

I have tried to emphasize that new types of research are
needed to find a way to make more interferon more economically
and a product perhaps not so heavily contaminated with the
qualities that feed the fears of those concerned with regulating
the public safety. An innovation that's badly needed and would

be a great help in studies concerned both with interferon
production and purification is a quantitative measurement of
interferon not based on a biological assay. Hopefully we will
see an immune assay, using either radioactive labels or one
of the enzyme linked methods emerge in the near future.

New drug delivery systems are constantly being devised.
Capsules and pumps for the constant slow release of a number
of substances have been developed. Better ways to deliver
interferon might enhance its possibilities and reduce the
amount needed. Work underway at Baylor University by Dr.
Robert Couch and his associates (personal communication)
suggest the manner in which interferon is administered intra-
nasally is important not only in determining the degree of
efficacy of interferon treatment but also in the amount of
interferon required. Studies such as this one may reopen the
possibility that interferon could be considered in the prophylaxis
of rhinovirus and influenza virus infections.

I would like to comment briefly on the need to standardize
the interferon preparations which are to be used in clinical
trials. Whether we are considering leukocyte interferon or
fibroblast interferon, their positions on SDS polyacrylamide
gels are readily determined. These gels which readily permit
the detection of a few micrograms of protein are simple and quick
to use. The beauty of their use is that they give one a picture
of contaminating proteins, also both their number and their
relative amounts. This might provide a simple tool for
standardizing interferons of a particular type especially when
different schemes are used for their partial purification.
Also, a before and after picture of the total protein profile
would detect the addition of foreign proteins that might be used
in purification methods.

As I close, I must conclude that interferon is where it
began. It has considerable theoretical appeal, it has shown
promise in osteogenic sarcoma and zoster and hepatitis B virus
infections as discussed earlier, but it has yet to become the
demonstrated treatment of choice for anything. I hope the
future will see interferon available at considerably less
expense so that enlarged clinical trials can be done. I hope
one or two of the ideas I have presented will be of help to
that end.

Acid treatment of crude
    interferon, 6
Actinomycin D, 38, 56, 67, 70,
    109, 127
Adenosine 3',5'-cyclic
    monophosphate (cAMP), 25,
    29
Affinity chromatography, 76
Amphotericin B, 104
Anchorage-dependent mammalian
    cells, production,
    microcarriers for, 15-23
Antifibroblast interferon, 76,
    78, 106-107
Antileukocyte interferon, 76,
    77, 78, 106-107
Antitumor activity, 159-174
    endotoxin and, 193-199
    in vitro, 161-165
    in vivo, 165-169
Antiviral activity, 85-99, 119-
    131, 135-136, 147, 194,
    201-211, 213-214
    double-stranded ribonucleic
        acid, 175-191
    leukocyte interferon, 78-82,
        90
    leukocyte vs. fibroblast
        interferon, 113-115
    mouse interferon, 87-90
Arbovirus, 203
Arenavirus, 203
Arrhenius theory, 147
Asparaginase, 215

Blood
    buffy coats of, 1, 2, 3, 4,
        61, 102, 113
    storage of, 1-3
Blood groups, interferon
    yield and, 41, 44
Bluetongue virus (BTV),
    interferon induction
    with, 37-53
Bovine interferon, 78, 79, 80,
    81, 82
Buffy coats, 1, 2, 3, 4, 61,
    102, 113
Burkitt's lymphoma, 159, 161,
    162, 165, 169, 215

Cancer therapy, interferon in,
    37, 38
Carbohydrates of interferon,
    215
Carboxymethylcellulose (CMC),
    17
Cell multiplication inhibitory
    (CMI) activity of inter-
    feron, 161-165, 169, 170
Cell plasma membrane and
    interferon therapy, 85-
    99
Cholera toxin, 34, 86, 87
Chronic hepatitis B, 202
Concanavalin A (Con A), 25,
    26, 27, 28, 29, 31
Conjunctivitis lignosa, 178,
    182, 187
Cycloheximide, 64, 67, 70, 109
Cytomegalovirus, 202

Cytopathic effect (CPE), 154

DEAE-dextran, 39
DEAE-Sephadex A50, 17, 18
Dibutyryl cAMP, 29-35
Dibutyryl guanosine 3',5'-
    cyclic monophosphate
    (cGMP), 31
Dulbecco's modified MEM, 39

Eagle's medium, 4-5, 124
Eagle's minimal essential
    medium (MEM), 104
Eagle's spinner medium, 125
Encephalomyocarditis virus,
    134, 176
Endotoxin, 194, 195
  antitumor activity of, 193-
    199
  interferon production and,
    193-199
Ethanol fractionation method, 5
Eye diseases
  double-stranded ribonucleic
    acid and, 175-191
  interferon in, 202

Fetal bovine serum (FBS), 39,
    104
Fibroblast interferon, 38, 45,
    61, 76, 78, 203, 214, 216
  embryo, 55-60
  freeze-dried, 144-147
  from human neonate foreskins,
    101-118
    properties, 113-115
  leukocyte vs., 62, 113-115,
    153-157
  lymphoblastoid, 61-74
  production, large-scale, 55-
    74, 101-118
  properties, 62
    thermal and vortical stabil-
    ity, 133-146
  sialic acid content, 153-
    157
Fibrosarcoma, 159, 164
Foreskin fibroblast cell strains
    for interferon, 101-118

Freeze-dried interferon, 144-
    147

Gangliosides, 85, 203
Gangliosides-interferon inter-
    action, 85-99
Gel filtration, 7
Gentamycin, 104
Glucose, 4
Glutamine, 4

Hemagglutinin (HA) assays,
    62
HEL 299 normal diploid human
    fibroblasts, 18
Hepatitis, 203
Hepatitis B, 214
Herpes simplex, 179, 181, 182,
    187
Herpes zoster, 179, 182, 187,
    188, 202
Hodgkin's disease, 168, 169,
    202
Human "agamma" serum, 4, 5
Human leukocyte interferon,
    102
  crude, 5
    acid treatment, 6
    purification, 5-9
    stability, 11-12
Human plasma protein fraction
    (HPPF), 39

Immune interferon, 25
Immune interferon induction,
    T cell, effect of low
    levels of cyclic ribo-
    nucleotides on, 25-35
Influenza virus, 202
Interferon. See also specific
    types
  administration, 202, 204-208
  as antiviral drug, 201-211
  carbohydrates of, 215
  cell multiplication inhibi-
    tory activity, 161-165
  for clinical use, 213-216
  endotoxin and, 193
  fibroblast vs. leukocyte, 62,

153-157
-ganglioside interaction, 85-
    99
induced by double-stranded
    ribonucleic acid, 176
neutralization tests, 106-107
production
    large-scale, 201, 203, 214-
    215
    in tumors, 193-199
    in vivo, tissue culture
    models of, 119-131
purification, 203, 209
response with cell types and
    sialic acid content, 153-
    157
side-effects, 202
standardization, 216
tissue-specificity, 165

Juvenile laryngeal papilloma,
    168-169

Keratoconjunctivitis, 182

L cell, 35, 39
Lassa fever, 203
Leukemia, 159
Leukocyte interferon, 113, 133,
    147, 160, 202, 203, 214,
    216
    antitumor activity, 159-174
    antiviral activities, 78-82
    blood group and, 41, 44
    in cancer therapy, 37, 38
    fibroblast vs., 62, 113-115,
        153-157
    immune, 25-35
    production
        from anchorage-dependent
        mammalian cells, 15-23
        blood, storage of, 1-3
        composition of medium, 4-5
        recovery of leukocytes, 4
        Sendai-virus, storage of,
            5
        virus-induced, 25-35
    production, large-scale
        in embryo fibroblast, 55-

    60
    virus-induced, 37-53
    properties
        acid treatment of crude,
            5-6
        antigenic, 75-78
        protein composition, 9, 10
        pyrogenicity, 1, 7-9
        stability, 11-12
    sterilization of, by membrane
        filtration, 9
    purification, 58
    in viral disease therapy, 37
    virus-type, 25, 28, 35
Leukocytes, 1, 2
    recovery of, 4
Limulus endotoxin pyrogen
    assay, 7, 8
Linear nonisothermal (LNS)
    accelerated storage test,
        147
Lymphoblast-derived interferon,
    55
Lymphoblastoid interferon, 203
    production of interferon by,
        61-74
Lymphoma, 159

Male genital herpes, 182, 187
Mammary carcinoma, 159, 164,
    169
Medium for cell cultivation
    composition, 4-5
Mengovirus, 125
Mercaptoethanol, 81
Methyl xanthine 3-isobutyl-
    1-methyl xanthine (IMX),
    34
Microcarriers for large-scale
    production of anchorage-
    dependent mammalian
    cells, 15-23
Migration inhibitory factor
    (MIF), 28
Minimal Essential Medium with
    Earle's salts (EMEM),
    39, 40
Mitogens, T cell, cyclic ribo-
    nucleotides and, 25-35

Moloney leukemia virus, 86, 94
Mouse fibroblast interferon,
    135
Mouse interferon, 78, 79, 80,
    81, 82, 86, 124, 147
  antiviral activity, 87-90
Multiple isothermal tests, 147
Multiple myeloma, 168, 169
Murine leukemia viruses, 86, 94,
    97

Namalva cell interferon, 55, 61
  production, large-scale, 61-
    74
Neomycin, 4
Neoplastic diseases and inter-
    feron therapy, 159-174
Neuraminidase, 153, 154, 156
Newcastle disease virus (NDV),
    38, 39, 41, 45, 49, 64,
    71, 121, 125, 126, 194
Nucleotide(s). See
    Ribonucleotides, cyclic

Osteogenic sarcoma, 214
Osteosarcoma, 159, 162, 164,
    165, 168, 169, 202

PFC response, 25-26, 29, 31,
    33, 34
  anti-SRBC, 26, 32, 34
Phytohemagglutinin (PHA), 86
Phytohemagglutinin P (PHA-P),
    25, 26, 27, 28, 29
Polyriboinosinate·polyribo-
    cytidylate (poly(I)·
    poly(C)), 38, 40, 41, 45,
    55, 56, 101, 102, 105, 109,
    113, 134, 154
Poly rI : poly rC, 67
Poly rl : rC + DEAE-Dextran, 64
Protein composition, 9, 10
Pyrogenicity, 1, 7-9

Rabies, 179, 203
Rhinoviruses, 202, 216
Ribonucleic acid, double-
    stranded, 202

antibody production and, 175
  in viral dermatoses and eye
    diseases, 175-191
Ribonucleotides, cyclic
  mitogen and, 25-35
  virus-induced production of
    interferon, 25-35
Rubella virus, 130, 202

SEA-induced immunosuppression,
    31, 34
Semliki forest virus, 62, 64
Sendai virus, 9, 38, 39, 40,
    41, 45, 49, 55, 62, 63,
    64, 67, 71, 154
  storage of, 5
Sindbis virus, 176
Soluble immune response
    suppressor (SIRS), 28
Staphylococcal enterotoxin A
    (SEA), 25, 27, 28, 29,
    31, 32, 33, 34
"Sterilin" Bulk Culture Vessel,
    57
Sterilization of interferon
    by membrane filtration,
    9
Swine interferon, 78, 80, 81,
    82

T cells, 28
T lymphocyte mitogens, 25
Thyrotropin (TSH), 86, 87
Tick-borne encephalitis
    (Hypr-strain), 176
Tricine, 4
Tumor necrosing factor (TNF),
    193, 197
Tumors
  endotoxin and, 193
  interferon production and,
    193-199

Verrucae vulgares et planae,
    182
Vesicular stomatitis virus
    (VSV), 154, 161, 194
Viral dermatoses, double-
    stranded ribonucleic

        acid and, 175–191
Viral diseases, therapy,
        interferon in, 37
Virions, 94, 130
Virus-type interferon, 25,
        28

Printed in the United States
by Baker & Taylor Publisher Services